**W9-BDT-082**

## s o l u t i o n s @ s y n g r e s s . c o m

With more than 1,500,000 copies of our MCSE, MCSD, CompTIA, and Cisco study guides in print, we continue to look for ways we can better serve the information needs of our readers. One way we do that is by listening.

Readers like yourself have been telling us they want an Internet-based service that would extend and enhance the value of our books. Based on reader feedback and our own strategic plan, we have created a Web site that we hope will exceed your expectations.

**Solutions@syngress.com** is an interactive treasure trove of useful information focusing on our book topics and related technologies. The site offers the following features:

- One-year warranty against content obsolescence due to vendor product upgrades. You can access online updates for any affected chapters.

- "Ask the Author" customer query forms that enable you to post questions to our authors and editors.

- Exclusive monthly mailings in which our experts provide answers to reader queries and clear explanations of complex material.

- Regularly updated links to sites specially selected by our editors for readers desiring additional reliable information on key topics.

Best of all, the book you're now holding is your key to this amazing site. Just go to **www.syngress.com/solutions**, and keep this book handy when you register to verify your purchase.

Thank you for giving us the opportunity to serve your needs. And be sure to let us know if there's anything else we can do to help you get the maximum value from your investment. We're listening.

## w w w . s y n g r e s s . c o m / s o l u t i o n s

SYN**G**RESS®

**1 YEAR UPGRADE**
BUYER PROTECTION PLAN

10 COOL

# LEGO®
# Mindstorms™

## ROBOTICS INVENTION
## SYSTEM 2™ PROJECTS

Jeff Elliott
Dean Hystad
Luke Ma
Dr. CS Soh
Rob Stehlik
Tonya L. Witherspoon

Technical Reviewers:
Mario Ferrari
Giulio Ferrari

| KEY | SERIAL NUMBER |
| --- | --- |
| 001 | M8KDR67VC2 |
| 002 | R45TVUH87H |
| 003 | Q2A4H7J9YB |
| 004 | Z4CX6BV44E |
| 005 | WSF6E6NKJ7 |
| 006 | ND56G7SW2S |
| 007 | ZAQ9HJH65D |
| 008 | VR54D7J8V2 |
| 009 | S5G7HF43CV |
| 010 | M39Z5BVY3X |

PUBLISHED BY
Syngress Publishing, Inc.
800 Hingham Street
Rockland, MA 02370

**10 Cool LEGO® MINDSTORMS™ Robotics Invention System 2™ Projects**

Printed in the United States of America

2 3 4 5 6 7 8 9 0

ISBN-13: 978-1-931836-61-6
ISBN-10: 1-931836-61-2

Technical Reviewers: Mario Ferrari and Giulio Ferrari       Cover Designer: Michael Kavish
Acquisitions Editor: Catherine B. Nolan                     Page Layout and Art by: Shannon Tozier
Copy Editor: Kate Glennon

Distributed by Publishers Group West in the United States and Jaguar Book Group in Canada.

# Acknowledgments

We would like to acknowledge the following people for their kindness and support in making this book possible.

A special thanks to Matt Gerber at Brickswest for his help and support for our books.

Karen Cross, Lance Tilford, Meaghan Cunningham, Kim Wylie, Harry Kirchner, Kevin Votel, Kent Anderson, Frida Yara, Jon Mayes, John Mesjak, Peg O'Donnell, Sandra Patterson, Betty Redmond, Roy Remer, Ron Shapiro, Patricia Kelly, Andrea Tetrick, Jennifer Pascal, Doug Reil, David Dahl, Janis Carpenter, and Susan Fryer of Publishers Group West for sharing their incredible marketing experience and expertise.

Duncan Enright, AnnHelen Lindeholm, David Burton, Febea Marinetti, and Rosie Moss of Elsevier Science for making certain that our vision remains worldwide in scope.

David Buckland, Wendi Wong, Daniel Loh, Marie Chieng, Lucy Chong, Leslie Lim, Audrey Gan, and Joseph Chan of Transquest Publishers for the enthusiasm with which they receive our books.

Kwon Sung June at Acorn Publishing for his support.

Jackie Gross, Gayle Voycey, Alexia Penny, Anik Robitaille, Craig Siddall, Darlene Morrow, Iolanda Miller, Jane Mackay, and Marie Skelly at Jackie Gross & Associates for all their help and enthusiasm representing our product in Canada.

Lois Fraser, Connie McMenemy, Shannon Russell, and the rest of the great folks at Jaguar Book Group for their help with distribution of Syngress books in Canada.

David Scott, Annette Scott, Delta Sams, Geoff Ebbs, Hedley Partis, and Tricia Herbert of Woodslane for distributing our books throughout Australia, New Zealand, Papua New Guinea, Fiji Tonga, Solomon Islands, and the Cook Islands.

# Contributors

**Rob Stehlik** is in his final year of studies in Mechanical Engineering at the University of Toronto. Rob has been avidly building mechanisms and robots with LEGO for three years. Fortunate enough to live in the Toronto area, he is an active member in the local LEGO enthusiasts' group, rtlToronto. Rob has participated in six LEGO robotics competitions organized by rtlToronto, and he credits much of his experience and inspiration to these events.

*Rob Stehlik is the creator of Robot 10: The RIS Turtle.*

**Dean Hystad** has spent much of the last 18 years building robots and testing equipment for MTS Systems Corporation in Minneapolis, MN. He was awakened from his dark ages–his non-LEGO years–when his loving and understanding wife gave him a Robotics Invention System (RIS) kit for Christmas three years ago. Since then, his obsession with LEGO (what else is there to do during a Minnesota winter?) has led to involvement in First LEGO League as a judge, mentor, and author of training materials.

*Dean Hystad is the creator of Robot 1: The Bug and Robot 8: The ULK.*

**Luke Ma** is a student at Brown University in Providence, RI. He is currently pursuing a bachelor's degree in Music and Computer Science. His main interest is in Music Theory, and thus he spends most of his time analyzing pieces of obscure classical music in even more obscure ways.

Luke has worked for Latitude Communications, Inc. as an engineering intern, helping the company develop and expand its Web-conferencing platform. He also has extensive experience in designing Web pages. He is fluent in C/C++, JavaScript, HTML/DHTML, Chinese, English, and hopefully French and German sometime in the future.

Luke would like to thank Catherine Nolan of Syngress for all her help and her courage to contract him as an author. Luke would also like to thank his parents for their support and his friends for putting up with him and making his life fun and enjoyable.

*Luke Ma is the creator of Robot 2: The Funky Chicken Techno-Walker, Robot 3: The Missle Turret, and Robot 4: The MINDSTORMS F1 Racer.*

**Jeff Elliott** is a jack-of-all-trades who divides his time between his work as a Software Development Consultant for Telepresence Systems, Inc., and his hobbies, which include creating LEGO models, scuba diving, rock climbing, and a host of other pastimes. Jeff lives in Toronto and is a founding member of the very active rtlToronto community. He has been creating LEGO robots since the early 1990s using the Dacta Control Lab, MINDSTORMS, and MicroScout products. He enjoys integrating LEGO robotics into his LEGO Train layouts, controlling switches, drawbridges, and car separators. His co-workers are becoming accustomed to the sight of a complex Train layout or LEGO robot gracing his office on a Monday morning.

*Jeff Elliott is the creator of Robot 9: The SpinnerBot.*


**Dr. Soh Chio Siong** (commonly known as **CSSoh** on the Internet) is a Public Health Physician who has a penchant for things scientific, mechanical, and electronic. Since he was a child, he has built crystal sets, microscopes, telescopes, steam engines, digital clocks, and computers, among other things.

Dr. Soh became interested in using LEGO as a tool for creative learning in 1998, with the purchase of some LEGO Dacta sets and, later on, the MINDSTORMS RIS set. He developed a special interest in pneumatics, particularly pneumatic engines, and is author of the world-renowned site on LEGO Pneumatics (www.geocities.com/cssoh1). He is an active member of the LUGNET community and has led many interesting discussion threads.

His current interest is the use of LEGO in the teaching of science and creativity. He thinks robotics should be the fifth R, after Reading, wRiting, aRithmetic, and computeR. He lives with his wife and daughter in Singapore.

Other LEGO claims to fame for Dr. Soh include: In September 1999, Dr. Soh's RCX Controlled Air Compressor Tester (www.lugnet.com/robotics/?n=7407) created quite a stir on the LUGNET Robotics discussion list.

CSSoh's LEGO Pneumatics Page (www.geocities.com/cssoh1) was voted LUGNET's Cool LEGO Site of the Week for January 9–15, 2000. This was the first site from Singapore to receive this recognition from LUGNET.

In June 2000, Dr. Soh, in collaboration with P.A. Rikvold and S. J. Mitchell of Florida State University, participated in a poster presentation at the Gordon Conference. The presentation, entitled "Teaching Physics with LEGO: From Steam Engines to Robots," can be viewed at www.physics.fsu.edu/users/rikvold/info/gordon00a.html.

*Dr. Soh is the creator of Robot 5: The Three-in-One Bot.*

**Tonya L. Witherspoon** is an Educational Technology Instructor at Wichita State University (WSU) in Wichita, KS. She teaches clay animation, multimedia production, Web design, and several robotics and programming courses using the LEGO MINDSTORMS RIS, Logo, Handy Crickets, and Roamer robots. She has co-authored several books on integrating technology into curriculum, speaks at state and national conferences on the subject, and teaches workshops and in-services for many schools in Kansas.

Tonya's interest in robotics peaked during Mindfest, a forum hosted by the Massachusetts Institute of Technology (MIT) in October 1999. She was inspired when Dr. Seymour Papert spoke about his work with MINDSTORMS and challenged everyone to encourage learning and find ways to spread knowledge in new and exciting ways. Since then, Tonya has received two grants that allowed her to give teachers in Kansas a MINDSTORMS RIS kit upon completion of a robotics workshop at Wichita State University. To date, she's given away over 75 RIS kits and helped many teachers find funding for complete classroom sets. She hosted a robotics summer camp this past summer for over 65 middle-school students; the camp also served as a practicum for teachers to learn how to use the MINDSTORMS RIS in their classrooms. In collaboration with WSU's College of Engineering, she has hosted two annual MINDSTORMS Robotics Challenges, events in which over 200 middle-school students from Kansas have competed in robotic challenges. The third annual MINDSTORMS Robotics Challenge will be hosted in March 2003 (http://education.wichita.edu/mindstorms).

Tonya's family consists of her husband, Steve, who is a teacher, and five school-age children: Andrew, Alex, Adam, Austin, and Madeline. She resides in Wichita, but lives in cyberspace.

*Tonya Witherspoon, in collaboration with her son, Alex, contributed Robot 6: The Aerial Tram and Robot 7: The LEGO Safe.*

**Alex Witherspoon** is a middle-school student in Wichita, KS. His brain is hard-wired for innovation; he has designed numerous creations on notebook paper since preschool. One of his first creations was a practical Midwestern solution: an explosive that would counteract and diffuse a tornado. Alex also designed a multi-level clubhouse, complete with a bed, television, computer, and a McDonalds on the lower level. He has made that clubhouse a reality in his backyard (minus the McDonalds). Alex presented his robot "Catapult Mania" at MIT's Mindfest when he was nine and broke the code to unlock the LEGO Knight's chain, which was a challenge posed to all Mindfest participants. His reward was to take home the four-foot LEGO Knight. Upon returning from Mindfest, Alex and his mother started a school-funded robotics club, at the invitation of Alex's elementary school principal.

The journey to MIT showed Alex that his type of creativity has ample application in our world, and has spawned different inventions using LEGOs and other materials to consummate the tenuous relationship between idea and reality. Alex has participated on robotics teams that have received the top prize for two years in a row at WSU's MINDSTORMS Robotics Challenge. This summer, he sent for a free patent attorney's kit.

*Alex Witherspoon, in collaboration with his mother, Tonya, contributed Robot 6: The Aerial Tram and Robot 7: The LEGO Safe.*

# Technical Reviewers

**Mario Ferrari** received his first LEGO box around 1964, when he was four-years-old. LEGO was his favorite toy for many years, until he thought he was too old to play with it. In 1998, the LEGO MINDSTORMS RIS set gave him reason to again have LEGO become his main addiction. Mario believes LEGO is the closest thing to the perfect toy. He is Managing Director at EDIS, a leader in finishing and packaging solutions and promotional packaging. The advent of the MINDSTORMS product line represented for him the perfect opportunity to combine his interest in IT and robotics with his passion for LEGO bricks. Mario has been an active member of the online MINDSTORMS community from the beginning and has pushed LEGO robotics to its limits. Mario holds a bachelor's degree in Business Administration from the University of Turin and has always nourished a strong interest for physics, mathematics, and computer science. He is fluent in many programming languages and his background includes positions as an IT Manager and as a Project Supervisor. With his brother Giulio Ferrari, Mario is the co-author of the highly successful book *Building Robots with LEGO MINDSTORMS* (Syngress Publishing, ISBN: 1-928994-67-9). Mario estimates he owns over 60,000 LEGO pieces. Mario works in Modena, Italy, where he lives with his wife, Anna, and his children, Sebastiano and Camilla.

**Giulio Ferrari** is a student in economics at the University of Modena and Reggio Emilia, where he also studied engineering. He is fond of computers and has developed utilities, entertainment software, and Web applications for several companies. Giulio discovered robotics in 1998, with the arrival of MINDSTORMS, and held an important place in the creation of the Italian LEGO community. He shares a love for LEGO bricks with his oldest brother, Mario, and a strong curiosity for the physical and mathematical sciences. Giulio also has a collection of 1200 dice, including odd-faced dice and game dice. Giulio has contributed to two other books for Syngress Publishing, *Building Robots with LEGO MINDSTORMS* (ISBN: 1-928994-67-9) and *Programming LEGO MINDSTORMS with Java* (ISBN: 1-928994-55-5). Guilio studies, works, and lives in Modena, Italy.

# About This Book

Each of the ten cool robots in this book is presented using a method that makes its construction as easy and intuitive as possible. Each chapter begins with a picture of the completed robot, accompanied by a brief introduction to the robot's history, its unique challenges and characteristics, as well as any concerns that the robot's creator wants you to be aware of during construction.

The instructions for building each robot are broken down into several sub-assemblies, which each consist of an integral structural component of the finished robot. (For example, the first robot presented in this book, the *Bug*, is broken down into six sub-assemblies: the Bumper, the Motor Mount, the Right Wheel, the Left Wheel, the Eye, and the Brow.) You will see a picture of each finished sub-assembly before you begin its construction.

You will be guided through the construction of each sub-assembly by following the individual building steps, beginning with Step 0. Each step shows you two important things–what parts you need, and what to do with them–by using two pictures. The *parts list* picture shows you which LEGO bricks you will need for that particular step, as well as the quantity of parts required, and the color of the parts (if necessary). Since this book is printed in black and white, we have used the following key to represent the colors:

- **B** Blue
- **G** Green
- **M** Magenta
- **LB** Light Blue

- **Y** Yellow
- **Ppl** Purple
- **TLG** Transparent Light Green
- **TY** Transparent Yellow

The *instructional* picture next to the parts list shows how those parts connect to one another. As the robot's construction progresses, it gets harder to see where parts get added, so you'll see we have made the parts that you add in each particular step *darker than* those added in previous steps. Many of the steps also have a few brief lines of text to more fully explain building procedures that may not be obvious from the pictures alone, or to discuss what role this step plays in the larger scheme of the robot's construction.

Once you have finished building all of the separate sub-assemblies, it's time to put them all together to complete the robot. The set of steps at the end of each chapter titled "Putting It All Together" walks you through the process of attaching together the sub-assemblies.

Throughout the chapters you will see three types of sidebars:

- **Bricks & Chips…** These sidebars explain key LEGO building concepts and terminology.
- **Developing & Deploying…** These sidebars explain why certain building techniques were used with a particular robot and what purpose they serve.

- **Inventing…** These sidebars offer suggestions for customizing the robots. Many of the robots in this book have alternate sets of building instructions that will radically change the overall function and performance of the finished robot. For example, the *Bug* robot has two variations on the standard set of building instructions presented in Chapter 1: a line following version of the robot and a version that incorporates a proximity sensor. Any alternate building instructions will be noted in these *Inventing* sidebars and can be downloaded from the Syngress Solutions Web site (www.syngress.com/solutions).

Building your robots is, or course, only half the fun! Getting them to run using the RCX brick is what distinguishes MINDSTORMS robots from ordinary models created with LEGO bricks. Some of the robots in this book will use the programs that come hard-wired into the RCX brick. Many of them will use unique programs that the authors have written specifically for their robots. Keep an eye out for the black and white *syngress.com* icons scattered throughout the book.

These icons alert you to the fact that there is code for this particular robot available for download from the Syngress Solutions Web site (www.syngress.com/solutions). The programs for the robots in this book are written in two of the most common programming languages used for LEGO MINDSTORMS:

- **RCX** LEGO's official programming language.
- **NQC** Standing for "Not Quite C," NQC is a programming language created by Dave Baum. Very similar in many ways to the C computer programming language, NQC is a text-based language that is more powerful and flexible than RCX.

For instruction on uploading these programs to your RCX brick, refer to the documentation that came with your LEGO MINDSTORMS RIS 2.0 kit.

The Syngress Solutions Web site (www.syngress.com/solutions) contains the code files and alternate building instructions for the robots found in *10 Cool LEGO Mindstorms Robotics Invention System 2.0 Projects: Amazing Projects You Can Build in Under an Hour.* The code files and alternate building instructions are located in a *BotXX* directory. For example, the files for Robot 5 are located in folder Bot05. Any further directory structure depends upon the specific files included for the robot in that particular chapter.

# Contents

The SpinnerBot, Robot 9

# Contents

# Robot 1

## The Bug

The Bug is a simple, differential-drive robot that was originally built to compete in a MINDSTORMS robot competition. The goal for the competition was to create a two-wheeled vehicle capable of navigating a figure-eight-shaped course. That original Bug was quite different from the one shown on the previous page. It didn't have a bumper, and the light sensor was mounted on a spar that extended from the front of the robot. The sensor was used to track a wide black line that ran down the center of the path the robots were to follow. Unfortunately, the Bug was disqualified when the contest director decided to allow only robots with a "bicycle-style" wheel configuration to enter the competition. As is the fate of most LEGO robots, the Bug was disassembled, its pieces returned to the parts bin.

The Bug was reincarnated almost a year later as part of a MINDSTORMS presentation given to generate interest in the FIRST LEGO League (FLL). I had purchased some Robotics Discovery Sets (RDSs) and wanted to use them as part of the demonstration. The RDS includes a blue programmable brick called the Scout. The Scout is programmed using the LCD and four buttons mounted on its faceplate; no external computer is required, and it only takes about five minutes of instruction before children can start writing their own robot control programs for it. A common Scout program for the Bug required it to wander around seeking light or darkness using the Scout's built-in light sensor. I added a forward-facing bumper to help the Bug navigate around obstacles, which you see in this version of the robot.

When doing presentations like the one at the FLL, I usually bring between eight and 10 robots: a variety of two-, four-, and six-legged walkers, Killough platforms, synchro drives, photo copiers, pick-and-place robots, and so on. But most people zoom right in on the Bug, perhaps because of the combination of its cute appearance, wobbly gait, and unusually inclined wheels. It's very common to hear the question, "Why did you put the wheels on that way?"

A two-wheeled robot like Bug is only stable if its center of gravity (COG) is lower than the axis of rotation of the wheels. If the COG is too high, the robot will tip over. The farther below the axis of rotation the COG is, the more stable the robot. The Bug is unusually stable for a two-wheeled robot. You can tip it more than 45 degrees forward or backward and it will return to an upright position. Adding off-center weight (such as the bumper and light sensor) has little effect on its attitude. It's even capable of climbing a gentle grade, or traversing small obstacles. The secret to the Bug's stability is the extreme camber of its wheels. When you look at the Bug from the side, the wheels appear to be elliptical (oval-shaped) instead of round. The flattened bottom of the ellipse closely matches the curvature of a circle with a much larger radius then that of the Bug's wheels. In fact, the 63 degrees of camber make the axis of rotation higher than that of a robot whose normal-oriented wheels (those oriented perpendicular to the ground) are twice the size of the Bug's wheels.

In addition to the base robot design for the Bug that is shown within the pages of this book, note that two alternative optimizations exist for the Bug: A line-following version and a proximity sensor version. Building instructions and programs for all versions of the bug are available for viewing and for download at the Syngress Solutions Web site (www.syngress.com/solutions).

# The Bumper

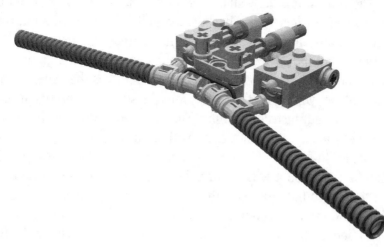

The Bug has "feelers" to help it investigate its surroundings. Pressure on the feelers tells the Bug it's time to stop, back up, and turn away. By having two feelers, the Bug can determine if the obstacle is on its left or right side.

### Bumper Step 0

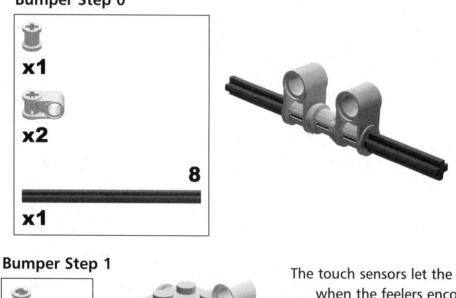

x1

x2

8

x1

### Bumper Step 1

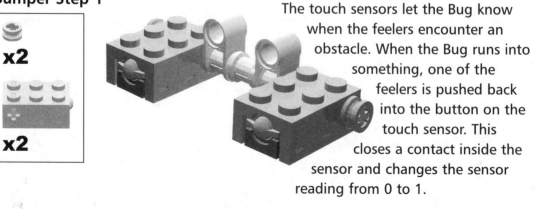

x2

x2

The touch sensors let the Bug know when the feelers encounter an obstacle. When the Bug runs into something, one of the feelers is pushed back into the button on the touch sensor. This closes a contact inside the sensor and changes the sensor reading from 0 to 1.

## Bricks & Chips...

## Bumper Design

The Bug uses a "normally open" style of bumper. The touch sensor button is pressed (that is, sensor contacts are closed) only during a collision.

Another popular design is the "normally closed" style of bumper. The normal state for these bumpers is with the feeler pressing against the touch sensor button. The feeler *releases* the button when a collision occurs.

## Bumper Step 2

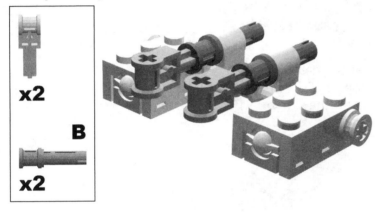

The TECHNIC pin with the stop bushing is used in this step to convert from a "pin type" connection to an "axle type" connection. Later on, the exposed pins are used to attach the bumper to the Bug's frame.

## Bumper Step 3

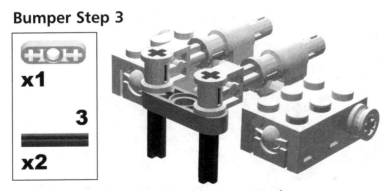

The 3L liftarm adds some much-needed rigidity to the bumper assembly.

## Bumper Step 4

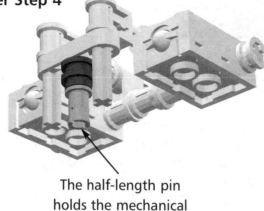

The round plates form a mechanical stop that holds the feelers in place. The plates are large enough to prevent the feelers from tangling, yet small enough not to impede the feeler's movement.

The half-length pin holds the mechanical stop in place.

### Bricks & Chips...

## Did You Know?

The short shaft on the half-length pin is the same size as the studs on top of LEGO bricks and plates. It can be used to put studs on the side of a TECHNIC beam.

## Bumper Step 5

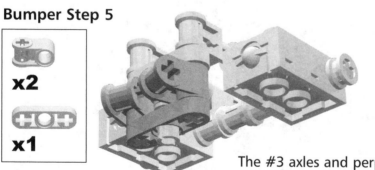

The #3 axles and perpendicular axle connectors make a hinge that allows the feelers to pivot freely.

## Bumper Step 6

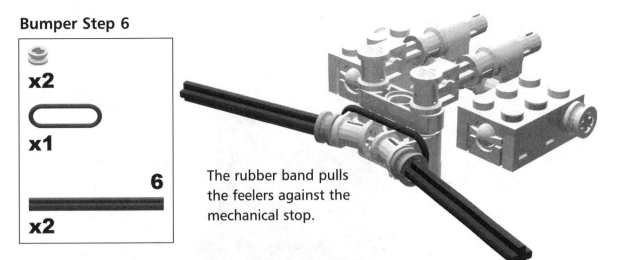

**x2**

**x1**

**6**

**x2**

The rubber band pulls
the feelers against the
mechanical stop.

### Bricks & Chips...

## Rubber Band Sizes

The rubber band must be small enough to hold the feelers against
the stop, otherwise the bumper might indicate a collision when
none has occurred. If the band is too small, the bumper will be
slow to respond to collisions. The white belt and the small black
rubber band are both just the right size.

## Bumper Step 7

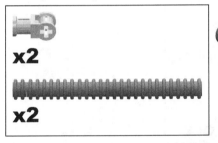

**x2**

**x2**

The ribbed hose
extends the reach of the
feelers. It also acts like a spring,
absorbing most of the shock from
collisions.

### Bricks & Chips...

## Problems with Hose Sizes

The ribbed hose does not appear to be made to the same exacting toler-
ances as other LEGO parts. Some fit very tightly when slid over an axle,
while others are loose. If you have a problem with the hose slipping off,
remove the hose and gently pinch the end (but not so hard as to kink it).
This will deform the hose a little and make the fit a bit tighter.

# The Motor Mount

The motor mount contains the motor and gearbox for driving the wheel. It also provides bracing to hold the whole robot together. Because there are two wheels, you will need to build two motor mounts.

### Motor Step 0

**2**
**x2**

**x4**

**3**
**x1**

**B**
**x1**

Use the blue angle connector with the stamped 2 on the side.

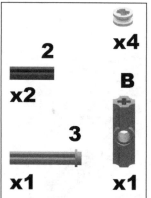

### Motor Step 1

**x2**

**x2**

**x1**

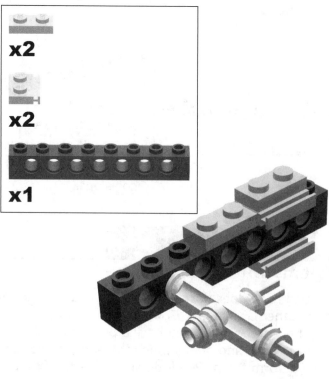

## Bricks & Chips...

# LEGO Terminology

The "stud" is the standard unit of length used when describing the size of LEGO parts. The TECHNIC beam in **Motor Step 1** is one stud wide by eight studs long. When we use terms like "1x8 beam" or "2x4 plate," the numbers refer to the width and length of the part measured in studs.

## Motor Step 2

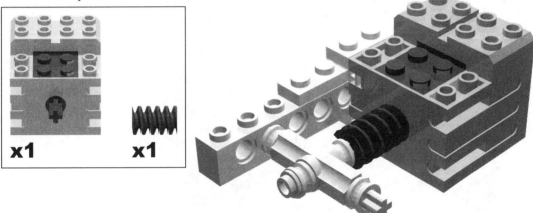

## Motor Step 3

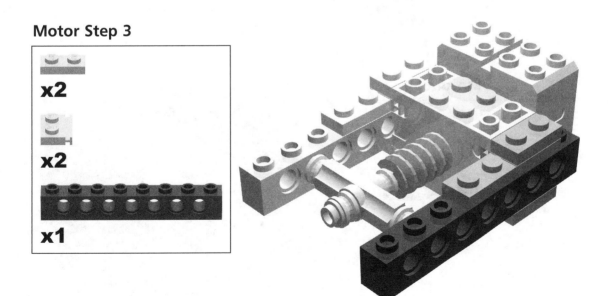

**Bricks & Chips...**

## Bricks, and Beams, and Chips... Oh, My!

TECHNIC beams (and LEGO bricks) are taller than they are wide. A beam is 1.2 studs in height, not counting the studs on top. Plates are one-third of the height of a brick, or 0.4 studs high.

It's important to note that TECHNIC plates differ from "standard" LEGO plates in that they have holes centered between the studs (LEGO plates do not have holes). The holes accept axles and connector pins, making the TECHNIC plates very useful. Use "standard" plates when you won't be using the through-holes, and save the TECHNIC plates for where they are needed.

### Motor Step 4

**x1**

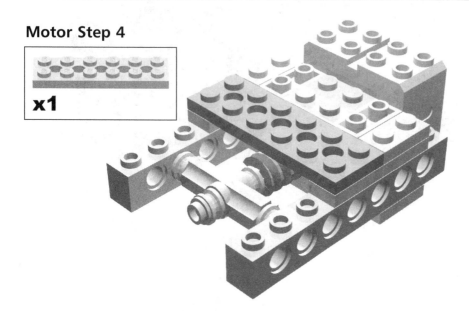

### Motor Step 5

**x4**

**x2**

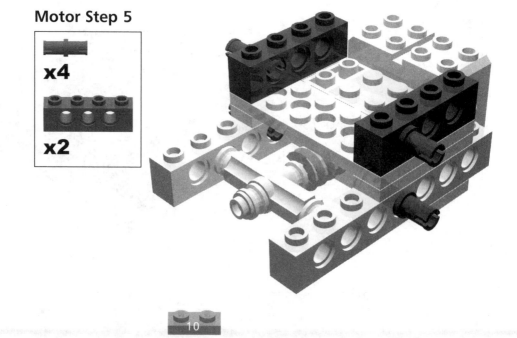

## Motor Step 6

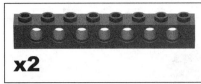

**x2**

Using the vertical beam to lock the 1x4 and 1x8 horizontal beams together is called *cross bracing*. Cross bracing makes this assembly very strong.

The height of the 1x4 beam (1.2 studs) and two plates (0.4 studs multiplied by two) places the pins two studs apart; this is just the right spacing to have them line up with holes in the vertical beam.

Knowing how to use combinations of plates and beams to make the holes in horizontal and vertical beams line up is an important LEGO building skill.

Remember you will need to build two of these.

# The Right Wheel

Now it's time to add the wheel and gearing to the motor mount.

## Right Wheel Step 0

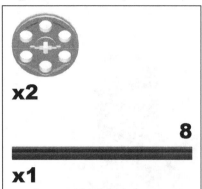

**x2**

8

**x1**

The pulley wheels act as a spacer to prevent the rubber tire from rubbing against the motor mount. They also act as a bearing to carry some of the weight of the robot. A bushing could do the same thing, but the pulley wheel spreads the weight out over a larger area. This makes it less likely that the forces will damage the robot.

## Right Wheel Step 1

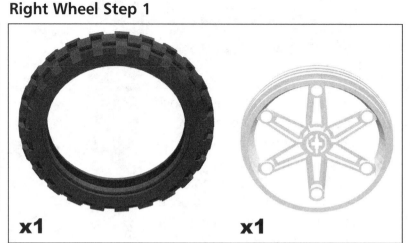

**x1**                    **x1**

## Right Wheel Step 2

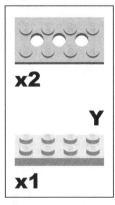

**x2**

**Y**

**x1**

The yellow plate is on the bottom and to the right of the gray TECHNIC plate. It's used here because there are not enough 2x4 TECHNIC plates. It also adds a nice splash of color to the otherwise gray-and-black assembly.

## Right Wheel Step 3

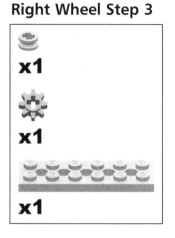

The 8t spur gear mates to the worm gear in the motor mount, providing an 8:1 gear reduction. Thus, the motor has to spin eight revolutions for the wheel to rotate just once. Gear reduction increases the torque output of the motor. The axle attached to the wheel may turn eight times slower than the motor shaft, but it also turns almost eight times stronger.

## Right Wheel Step 4

Locate one of the motor mount sub-assemblies that you built earlier. Slide the motor mount sub-assembly onto the axle, and secure the mount onto the axle with the two stop bushings.

# The Left Wheel

The left wheel is a mirror image of the right wheel.

## Left Wheel Step 0

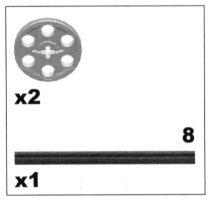

**x2**

8

**x1**

Begin the wheel assembly just as you did in **Right Wheel Step 0**.

## Left Wheel Step 1

**x1**　　　　　**x1**

Continue the wheel assembly just as you did in **Right Wheel Step 1**.

## Left Wheel Step 2

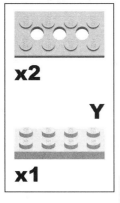

**x2**

**Y**

**x1**

For the left wheel, the yellow plate is on the bottom and to the *left* of the gray 2x4 TECHNIC plate.

## Left Wheel Step 3

**x1**

**x1**

**x1**

## Left Wheel Step 4

**x2**

Locate the second motor mount sub-assembly and slide it onto the axle. The two stop bushings help hold the wheel assembly and the motor mount together. This is the weakest connection in the Bug's design. After extended use, the wheel assembly and motor mount may begin to separate. If this occurs, correct the separation by snapping the bottom plates back in place and sliding the bushings down the axle until they contact the upper plate.

# The Eye

The Bug's eyes are not just decorative elements. They are part of a quick-disconnect fastener that makes battery changes fast and easy. You will need to build two eyes for the Bug.

## Eye Step 0

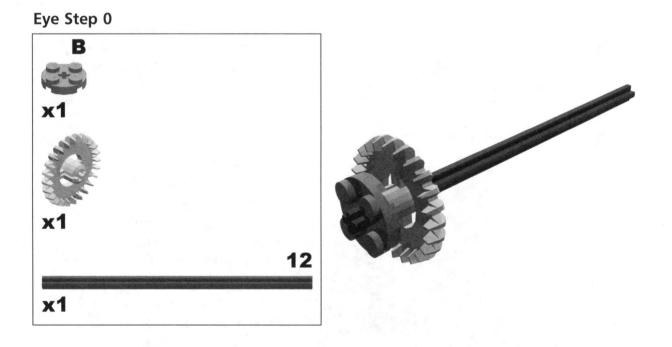

# The Brow

Originally the Bug had two long yellow antennae made out of the skinny flex hose. When I lost one of the hoses, I replaced them both with axles. One day, while playing with my robots, I realized that the antennae worked much better as eyebrows. By adjusting the angle of the eyebrows, you can make the Bug look angry, mildly interested, or really surprised.

   The eyebrows are purely decorative elements. Modify them as you see fit. The Bug requires two eyebrows.

**Brow Step 0**

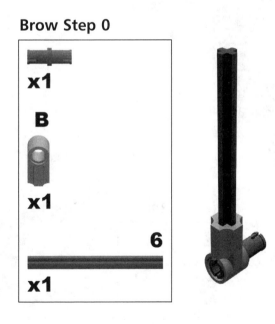

# The Pincer

The pincers are another purely decorative element, but they add a lot of character. The Bug requires two pincers.

### Pincer Step 0

x2

x1

Y

x1

Remember to build two pincers!

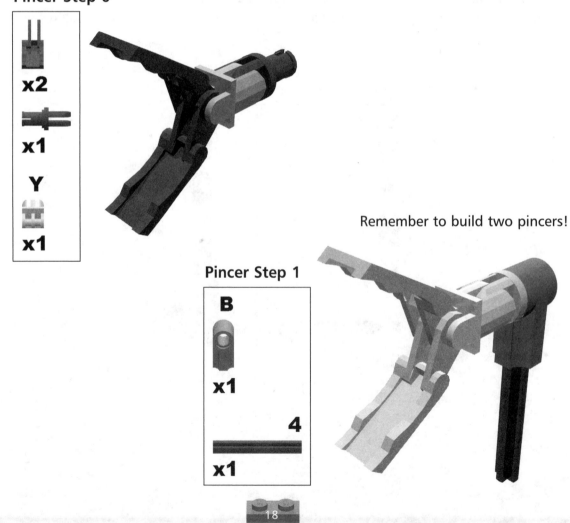

### Pincer Step 1

**B**

x1

4

x1

# Putting It All Together

Now that you've created the wheels, the eyes, the brows, the pincers, and the bumper, it's now time to put them all together!

### Final Step 0

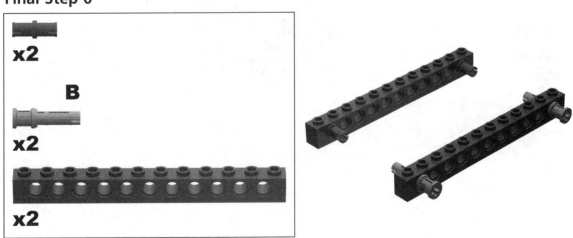

## Final Step 1

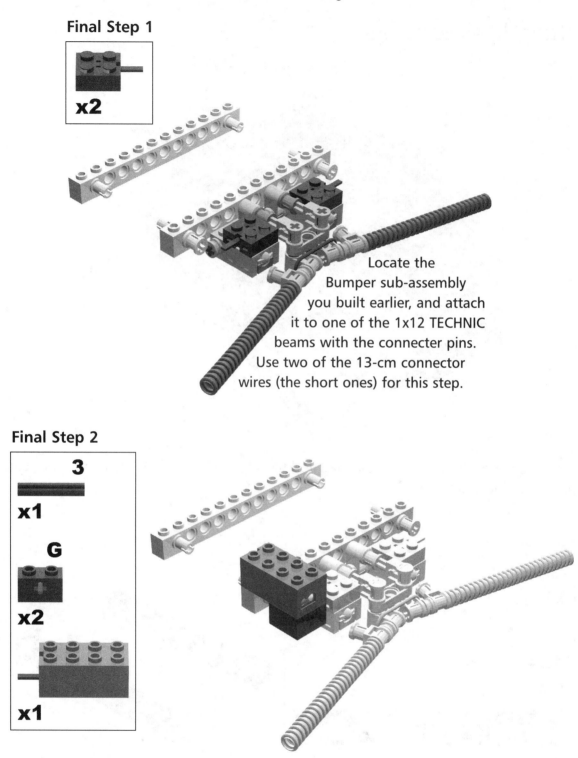

**x2**

Locate the
Bumper sub-assembly
you built earlier, and attach
it to one of the 1x12 TECHNIC
beams with the connecter pins.
Use two of the 13-cm connector
wires (the short ones) for this step.

## Final Step 2

**3**

**x1**

**G**

**x2**

**x1**

## Final Step 3

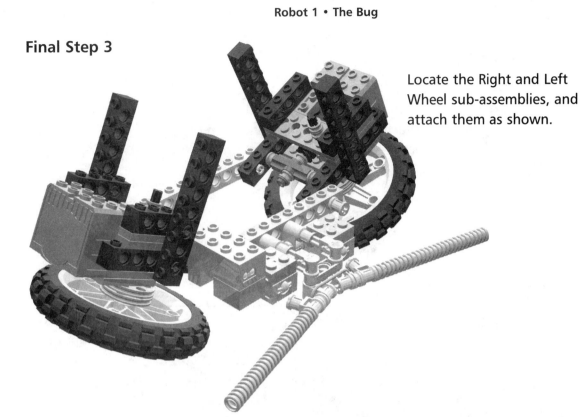

Locate the Right and Left Wheel sub-assemblies, and attach them as shown.

## Final Step 4

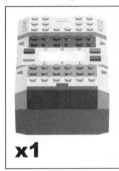

**x1**

Position the RCX so it rests against the side of the light sensor. The front of the RCX is aligned with the front of the light sensor.

## Final Step 5

**x3**

Route the wires under the 1x12 beam attached to the bumper and then up between the motor mount and the RCX.

The light sensor is attached to sensor input 2.

Attach the bumper's left touch sensor to Sensor Input 1.

Attach the bumper's right touch sensor to Sensor Input 3.

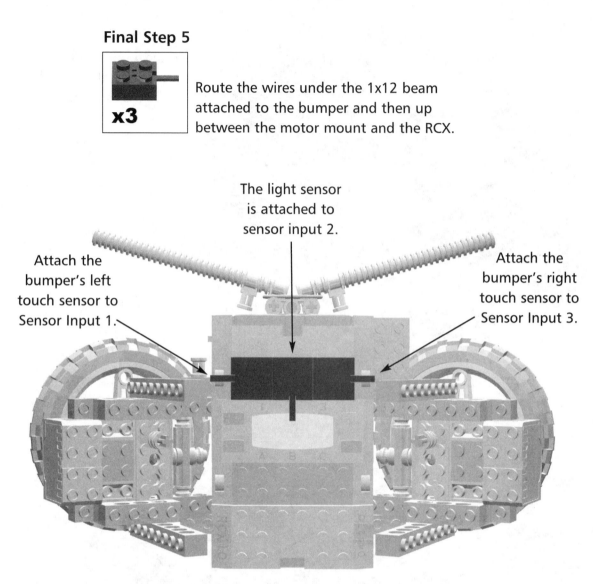

### Final Step 6

**x2**

Use the 13-cm connector wires to attach the motor to the RCX. Be sure to position the connectors as shown.

The Bug should travel forward when both motors are driven forward.

The left motor wire connects to Output A.

The right motor wire connects to Output C.

# Final Step 7

x4

x1

# Final Step 8

x2

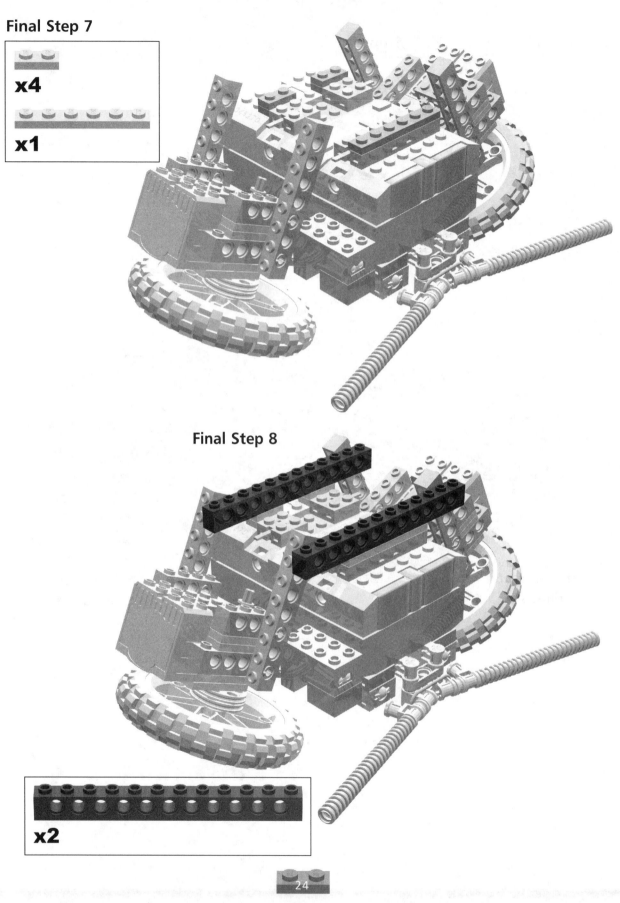

## Final Step 9

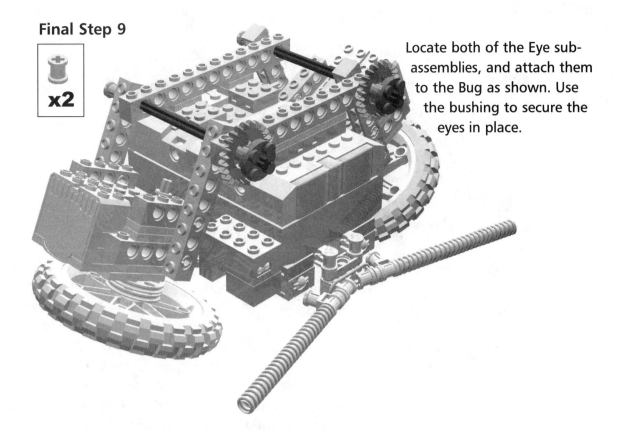

Locate both of the Eye sub-assemblies, and attach them to the Bug as shown. Use the bushing to secure the eyes in place.

## Bricks & Chips...

### Changing Batteries

To disassemble the Bug for a battery change, remove one of the eyes and separate the RCX from the dark gray battery case. With a little practice, you should be able to replace all six batteries without removing a single brick, plate, or wire.

## Final Step 10

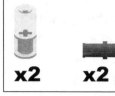

**x2**   **x2**

With the addition of the pincers and eyebrows to the robot as shown, the Bug is complete! Load and run the *cockroach* program to make the Bug seek out the dark corners of a room. Run the *moth* program, and the Bug will be drawn to bright lights. Shine a flashlight at the Bug, and he'll follow you around.

SYNGRESS
syngress.com

## Inventing...
## Programs and Add-ons

The programs for the Bug are located on the Syngress Solutions Web site (www.syngress.com/solutions). There are also two variations of the Bug available at this site: A line-following version and a version that incorporates a proximity sensor!

# Robot 2

## Funky Chicken Techno-Walker

Most common forms of transportation rely on wheels to move. However, most living beings do not have wheels; we use legs! Naturally, as we design robots to be some limited imitations of ourselves, we would like some of them to be able to "walk." Walking has many advantages over wheels: A walker can adjust to many different terrains, climb up stairs, and climb down stairs, just to name a few. But even the most advanced robotic walker in the world can only barely accomplish these feats that we as humans perform every day without thinking.

This is a simple robot that mimics the basic patterns of walking. This model is a walker, but not a complex one. It cannot walk up stairs, handle terrains, or even turn. Robots that can do those things require a tremendous number of motors, joints, and complex programs to control it. It has two legs, each of which make the correct walking motions. In order to achieve the correct timing for both the legs, the two legs are rotationally offset and joined by one axle. This allows one motor to drive two legs creating two motions! This model relies on completely mechanical solutions to solve the walking problem. But it *can* walk forward and does so with a minimum of parts and requirements. There is no programming needed to get this model to walk correctly.

Since we are using only one motor to drive the entire model, the model cannot be too heavy. The heaviest elements of any robot in real life or LEGO are usually the batteries. We can solve this problem easily here by not attaching the RCX to the body of the model. When you are finished with the model, simply connect the wires from the motors to the RCX brick to make the model walk. Be sure to use long connecting wires from the motors to the RCX brick. Short wires will work but long wires will allow your walker to walk more freely.

# The Top of the Body

You will start building the walker by constructing three sub-assemblies that make up its body, the top and the bottom sub-assemblies, and the motor. We will start by assembling the top of the body.

### Top of Body Step 0

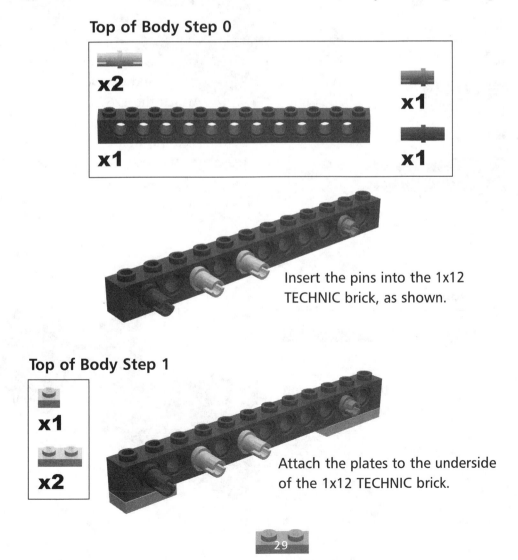

Insert the pins into the 1x12 TECHNIC brick, as shown.

### Top of Body Step 1

Attach the plates to the underside of the 1x12 TECHNIC brick.

## Top of Body Step 2

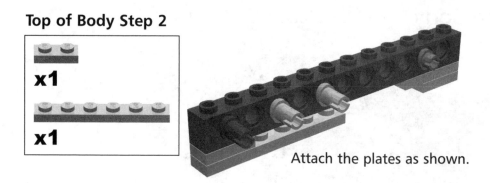

**x1**

**x1**

Attach the plates as shown.

## Top of Body Step 3

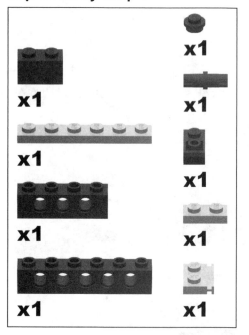

**x1**

**x1**

**x1**

**x1**

**x1**

**x1**

**x1**

**x1**

Attach plates and pins as shown. Attach the 1x6 TECHNIC brick and the 1x1 round plate before securing them with the 1x2 plate. The 1x2 plate with door rail should be on top of the 1x4 TECHNIC brick.

## Top of Body Step 4

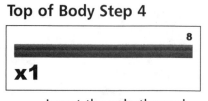

8

**x1**

Insert the axle through the brick, as shown.

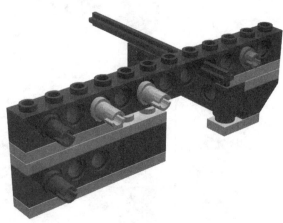

## Top of Body Step 5

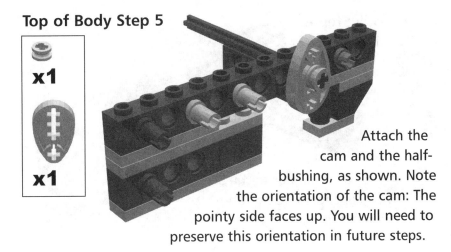

**x1**

**x1**

Attach the cam and the half-bushing, as shown. Note the orientation of the cam: The pointy side faces up. You will need to preserve this orientation in future steps.

## Top of Body Step 6

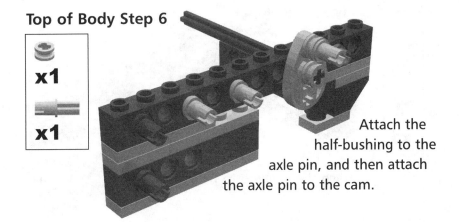

**x1**

**x1**

Attach the half-bushing to the axle pin, and then attach the axle pin to the cam.

## Top of Body Step 7

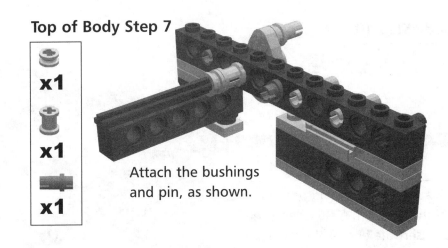

**x1**

**x1**

**x1**

Attach the bushings and pin, as shown.

## Top of Body Step 8

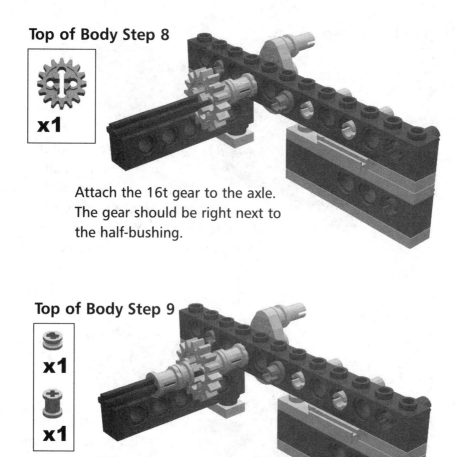

Attach the 16t gear to the axle.
The gear should be right next to
the half-bushing.

## Top of Body Step 9

**x1**

**x1**

Attach the bushing to the axle.

## Top of Body Step 10

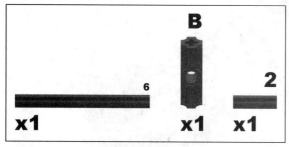

**B**

**6**

**2**

**x1**   **x1**   **x1**

Attach the #2 axle to one end of the angle
connector, and attach the #6 axle to the
other end. Insert the end of the #6 axle
into the middle of the 1x6 TECHNIC brick.

## Top of Body Step 11

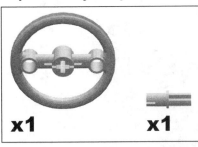

**x1**     **x1**

Attach the large pulley to the axle, and insert the axle pin into the pulley. It is often advantageous to extend a controlling axle so that you can control motion manually. In Top of Body Step 11, the pulley you attach to the axle allows you to turn the pulley and test the walker manually before you run the motors.

## Top of Body Step 12

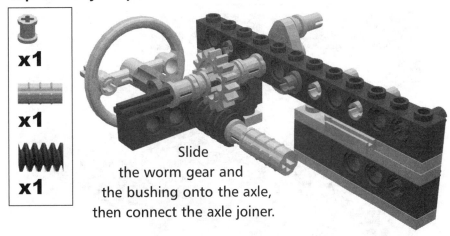

**x1**

**x1**

**x1**

Slide
the worm gear and
the bushing onto the axle,
then connect the axle joiner.

---

## Bricks & Chips...

## Worm Gears

Worm gears have a unique property among gears in LEGO—they slide. When you want a worm gear to actually drive another gear, you don't want the worm gear to slide and disengage. In these circumstances, always make sure to brace the worm gear with bushings, so that the worm gear does not slip.

---

## Top of Body Step 13

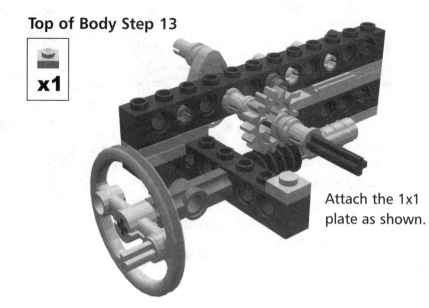

Attach the 1x1 plate as shown.

## Top of Body Step 14

Slide the TECHNIC brick through the axle, and attach it to the 1x1 plate from Top of Body Step 13.

## Top of Body Step 15

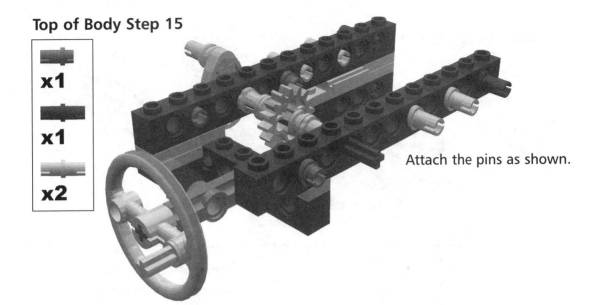

Attach the pins as shown.

## Top of Body Step 16

Adjust the assembly so that the cam from Top of Body Step 5 has its pointy end facing upward. Then attach the cam in this step to the axle, with its pointy end facing downward. Secure the cam with the half-bushing.

## Top of Body Step 17

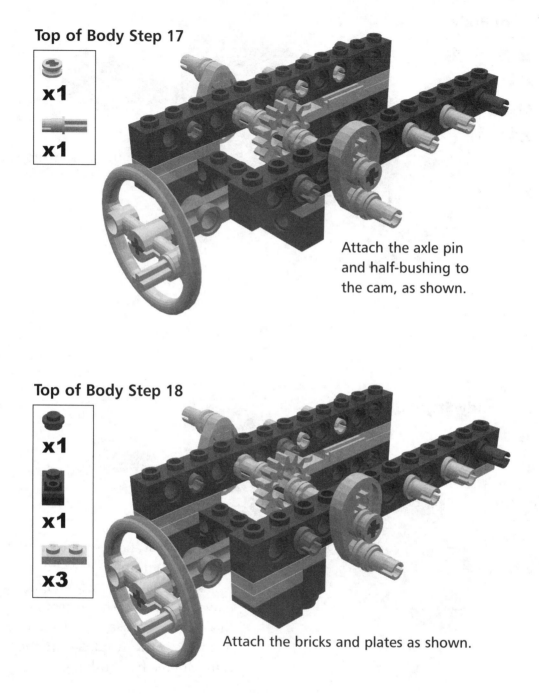

x1

x1

Attach the axle pin and half-bushing to the cam, as shown.

## Top of Body Step 18

x1

x1

x3

Attach the bricks and plates as shown.

## Top of Body Step 19

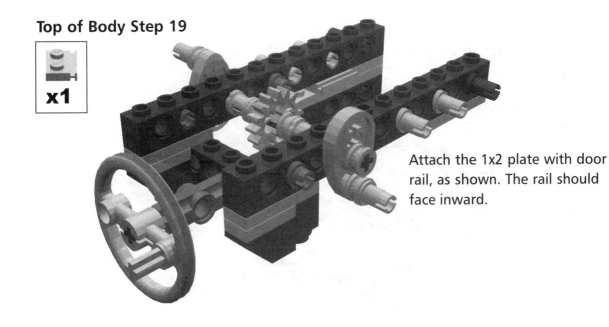

Attach the 1x2 plate with door rail, as shown. The rail should face inward.

## Top of Body Step 20

**x2**

**x1**

**x1**

**x1**

**x1**

**x1**

**x1**

First, attach all the plates. Then insert the pin with friction into the hole in the 1x4 TECHNIC brick that is furthest away from the gears. Finally, stabilize the assembly by attaching the 2x4 L liftarm as shown.

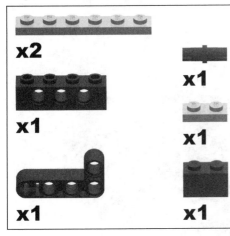

## Top of Body Step 21

**x1**

Attach the pin
as shown.

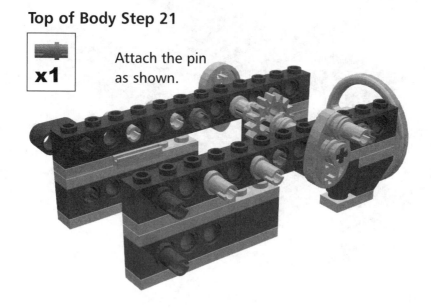

## Top of Body Step 22

**x1**

Attach the liftarm as shown.

# The Bottom of the Body

This sub-assembly creates the bottom of the body. The top and bottom are separated so that they are easier to build and then put together.

### Bottom of Body Step 0

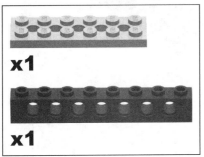

**x1**

**x1**

Attach the brick to the plate as shown.

### Bottom of Body Step 1

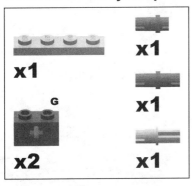

**x1**

**G**

**x2**

**x1**

**x1**

**x1**

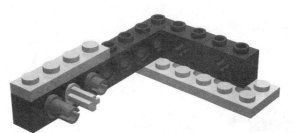

Attach the bricks, plates, and pins, as shown.

## Bottom of Body Step 2

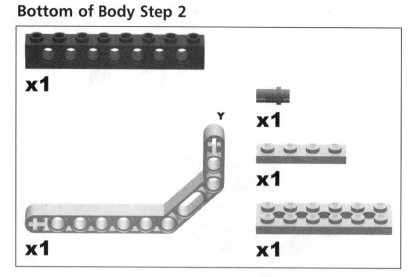

**x1**

Y

**x1**

**x1**

**x1**

**x1**

Attach the liftarm, brick, plates, and pins, as shown.

## Bottom of Body Step 3

**x2**

**x1**

**x1**

Attach the pins as shown.

## Bottom of Body Step 4

Y

**x1**

Attach the liftarm as shown.

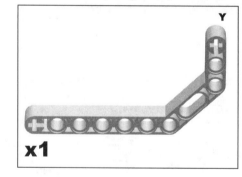

# The Walk Motor

This is the motor that will power the walker. We just need to add one little piece to the motor to make it more stable.

## Walk Motor Step 0

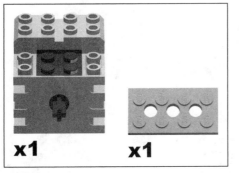

x1     x1

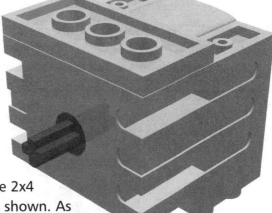

Flip the motor upside down. Add the 2x4 plate to the bottom of the motor as shown. As will be evident in future steps, the plate on the bottom of this motor will sit on top of the liftarms. This helps support the motor more and prevents undesirable vertical movement of the motor.

# Assembling the Body

Now you will put the entire body together and brace the components so that they are stable.

## Body Step 0

Get out the sub-assembly for the top of the body. Attach the sub-assembly for the bottom of the body to the sub-assembly for the top of the body.

## Body Step 1

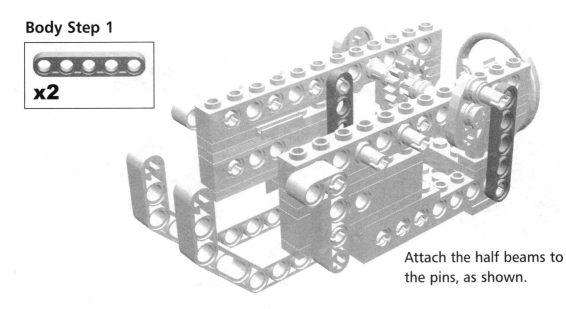

**x2**

Attach the half beams to the pins, as shown.

## Body Step 2

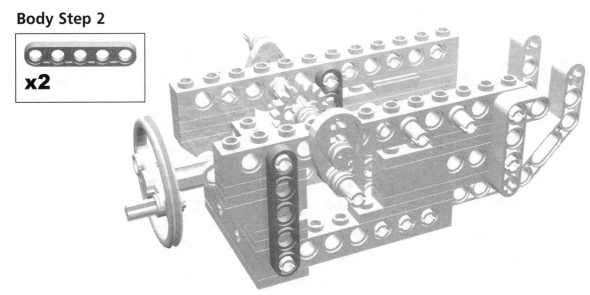

**x2**

Attach the half beams to the pins, as shown.

## Bricks & Chips...

### Bracing Beams

Beams are often a good way to brace a structure. They are thinner than bricks and are rounded, which means they fit into tight spaces better than bricks.

## Body Step 3

**x1**

**x1**

Slide the motor sub-assembly into the body. The slots on the side of the motor should be aligned with the 1x2 plates with door rail. Attach the wire brick to the top of the motor. Make sure the wires run along the groove on the top of the motor and out the back.

## Body Step 4

**x1**

Attach the plate as shown; this secures the motor. Make sure the wire is hanging out the back of the motor (not shown).

### Design & Planning...

## Planning Wire Placement

The wire needs to face out the back of the motor for two reasons: It will be easier to connect the wires to the RCX, and you would not want the wire to get caught in the gears.

# The Right Leg

Now that you have the body to power your walker, you will create the legs that will enable it to walk. We start with the right leg.

### Right Leg Step 0

### Right Leg Step 1

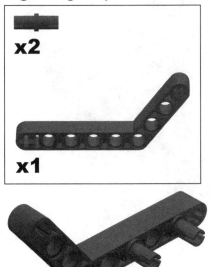

Take the lift arm and position it with the long end flat on the ground. Insert the pins as shown.

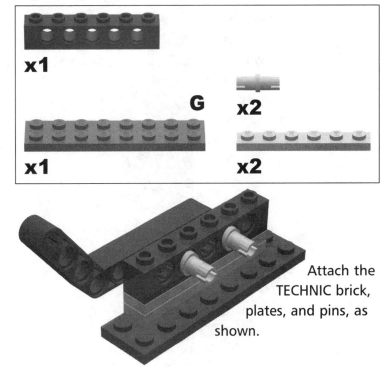

Attach the TECHNIC brick, plates, and pins, as shown.

### Right Leg Step 2

G

**x1**

Y

**x1**

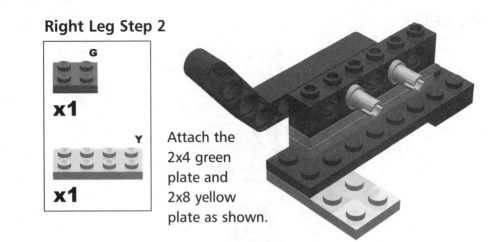

Attach the 2x4 green plate and 2x8 yellow plate as shown.

### Right Leg Step 3

**x1**

**x1**

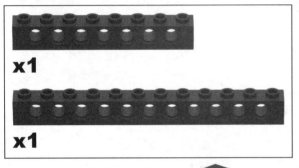

Attach the two bricks as shown. Make sure that the studded sides of the bricks face outward.

### Right Leg Step 4

**x2**

Insert the two axle pins into the bricks.

### Right Leg Step 5

**x1**

Connect the axle pins and bricks together using a 1x3 beam.

### Right Leg Step 6

**x2**

Turn your leg around and insert the pins as shown.

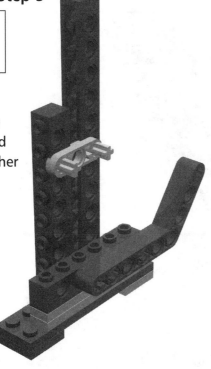

### Right Leg Step 7

**x2**

Attach the two bricks as shown. Make sure that the studded sides of the bricks face outward.

# The Left Leg

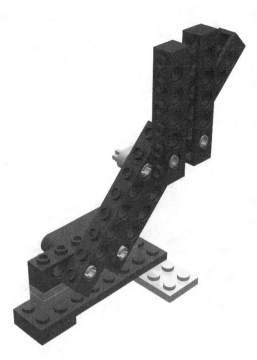

Now you will make the left leg, which is a mirror image of the right leg.

### Left Leg Step 0

x2

x1

Attach the pins to
the liftarm, as
shown.

### Left Leg Step 1

x2

x1

Attach the brick
and pins, as shown.

## Left Leg Step 2

**G**

x1

**Y**

x1

x2

**G**

x1

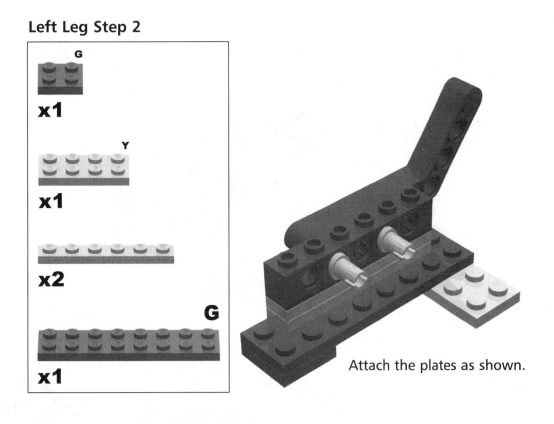

Attach the plates as shown.

## Left Leg Step 3

x1

x1

Attach the two bricks as shown.
Make sure that the studded
sides of the bricks face outward.

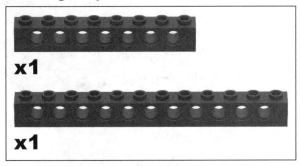

### Left Leg Step 4

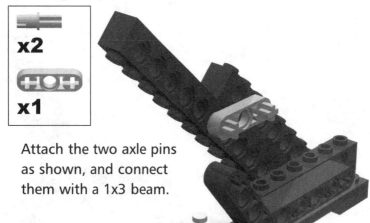

x2

x1

Attach the two axle pins
as shown, and connect
them with a 1x3 beam.

### Left Leg Step 5

x2

Turn your leg
around, and
insert the pins
as shown.

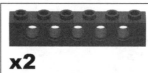

### Left Leg Step 6

x2

Connect the two
bricks as shown.
Make sure that
the studded sides
of the bricks
face outward.

# The Body with Legs

Now that you have built the body and the legs, you can put the walker together and make it walk!

### Body with Legs Step 0

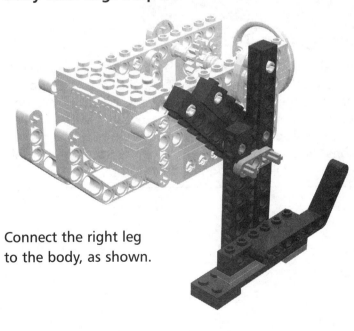

Connect the right leg
to the body, as shown.

## Body with Legs Step 1

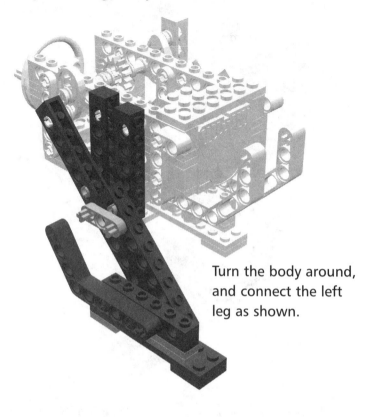

Turn the body around, and connect the left leg as shown.

## Inventing...
## Analyzing Your Walker

This model is made so that it can perform the correct walking motions on only one motor. An essential part of this walking mechanism is the opposite-pointing cams. What if the cams were not pointing in opposite directions? How would the walker walk if the cams were pointing in the same direction? Or in perpendicular directions? You can test these motions with the hand crank at the back of the model or by running the motor.

# The Head Turntable

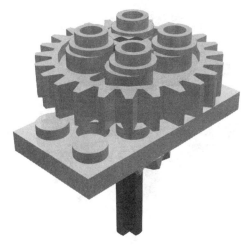

You will now add a spinning head to your walker. We start with a small sub-assembly that will enable the head to endlessly spin on the chicken neck.

### Head Turntable Step 0

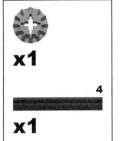

**x1**

**x1** 4

Insert the axle through the bevel gear as shown.

### Head Turntable Step 1

**x1**

Slide a 2x4 TECHNIC plate through the axle from the top and rest it on the bevel gear.

### Head Turntable Step 2

**x1**

Attach the 24t gear to the top of the axle. The 24t gear should barely clear the studs on the 2x4 TECHNIC plate beneath it. The top of the axle should also be flush with the top of the gear.

### Head Turntable Step 3

**x4**

Insert the pins into the holes in the gear.

## The Head

Here you will make the entire spinning head.

### Head Step 0

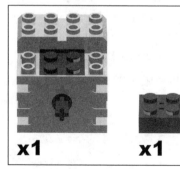

Take a motor and connect the wire brick to it. Make sure the wire runs out the back of the motor and along the groove at the top of the motor.

## Head Step 1

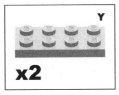

**x2**

Attach the two yellow 2x4 plates as shown. Make sure the wire goes out the back of the motor.

## Head Step 2

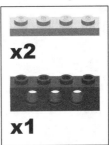

**x2**

**x1**

Attach the brick and the 1x4 plate as shown.

## Head Step 3

**x1**

**3**

**x1**

Attach the axle and half-bushing as shown.

## Head Step 4

**x1**

Attach the bevel gear as shown. Make sure the end of the axle is flush with the bevel gear.

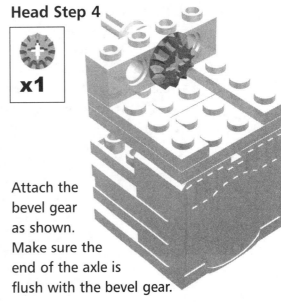

## Head Step 5

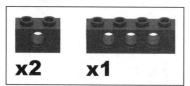

**x2**   **x1**

Attach the bricks as shown.

## Head Step 6

**x2**

Attach the plates as shown.

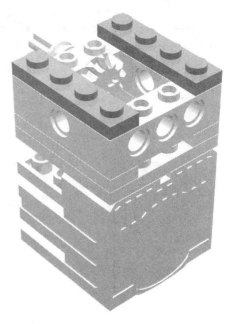

## Head Step 7

Attach the turntable sub-assembly, as shown, by resting the 2x4 plate across the top of the 1x4 plates from Head Step 6.

## Head Step 8

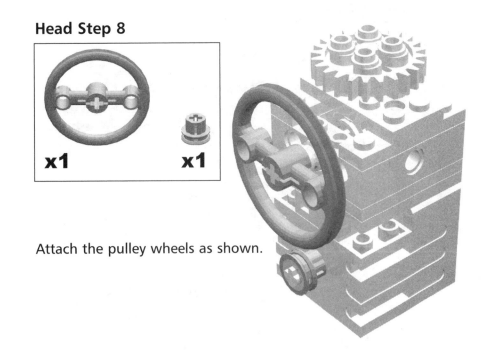

Attach the pulley wheels as shown.

## Head Step 9

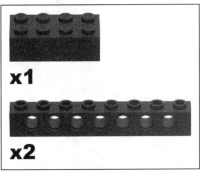

Attach the brick as shown. The 2x4 brick should be attached to the gear by the pins on the gear.

### Head Step 10

**x2**

Use two white radar dishes as eyes.

### Head Step 11

**3**

**x2**

**B**

**x2**

Stick the two blue pins with bushing through the two end holes of the bricks, as shown. Then insert two #3 axles into the pins.

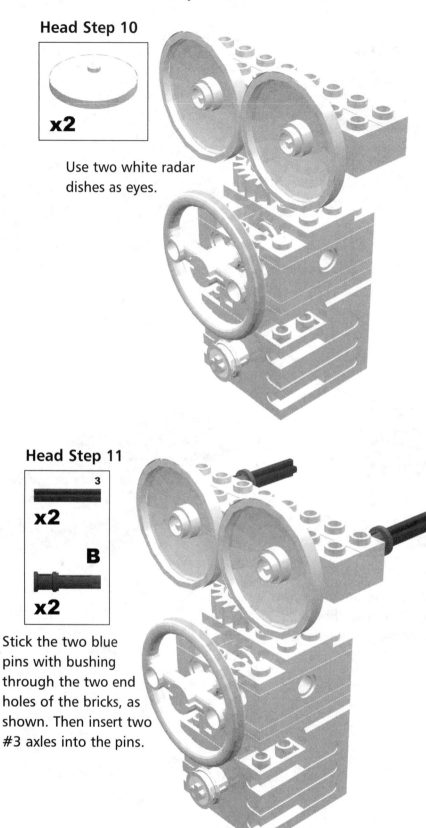

## Head Step 12

Give our head some odd antennas by attaching a tube (in your choice of color) to each axle from Head Step 11. Then attach a yellow rubber band to both the large and small pulley wheels.

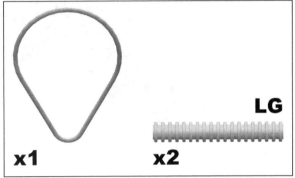

**LG**

x1    x2

## Gears versus Pulleys

A geartrain is not always the best solution for transfer of power. In this robot, we achieve a transfer of power from the motor to the head by use of a pulley system. The pulley system also has a natural "clutch" system—when the motor experiences a certain amount of resistance, it will still spin but the rubber band will slip. This prevents the motor from stalling, which is a good thing since a motor will drain a lot more power from your batteries when it is stalled than when it is functioning normally.

# Putting It All Together

You now have a body with walking legs and a head. It's time to finish your creation. Connect the body and legs sub-assembly to the head sub-assembly as shown, and you're done!

## Testing Your Walker

Before you run the walking motor to make this robot walk, make sure you know which direction the motor should turn to make the walker walk forward. You can test this with the hand crank at the back of the model. If you get the wrong direction for the motor, you'll see the model trying to walk backward.

# Robot 3

## Missile Turret

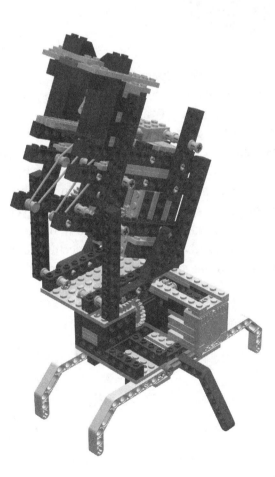

The missile turret is an extremely versatile defense system employed every-where from Earth to the outer reaches of the Alpha Centauri system. Its ability to mark targets in all directions is its most valued asset.

This is a LEGO model of a standard missile turret. While it may not keep away unwanted aliens, you could use it to protect your room from unwanted visitors. The model features a working firing mechanism with a hammer that cocks and has an automatic firing action. The Turret sits on a stand that fea-tures a turntable so that you can fire projectiles in any direction you want. You can also fire any ammunition you want, so long as it fits into the loading bay of the firing mechanism (1x1 or 1x2 bricks are ideal projectiles). Beware though–using this model is a fun but very easy way to lose pieces! Watch where you fire your shots!

# The Bracers

You will build the Turret in two parts: the stand and the firing mechanism. You will start with the stand. These bracers give the stand horizontal stability so that the stand will not fall over from the weight of the firing mechanism on top. You will need to build **two** of these.

## Bracers Step 0

B

x1

3

x1

Insert the axle into the blue angle connector.

## Bracers Step 1

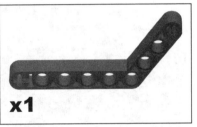

x1

Slide the liftarm onto the axle. Make sure you connect the shorter end of the liftarm to the axle.

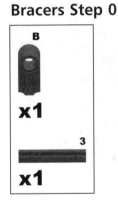

## Bracers Step 2

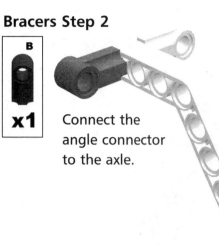

Connect the angle connector to the axle.

You need to build two bracers, so repeat Bracer Step 0 through Bracer Step 2.

# The Stand Motor

This is the motor assembly that will spin the firing mechanism. This will allow your turret to shoot in many different directions.

## Stand Motor Step 0

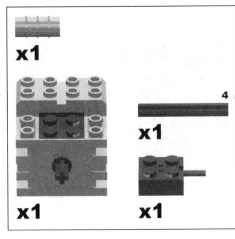

Connect the axle joiner to the motor and then insert the axle into the axle joiner. Make sure the wire of the wire connector brick sits in the groove at the top of the motor—you don't want the wire to get caught up in the gears.

## Stand Motor Step 1

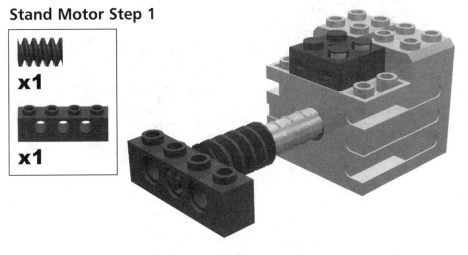

Slide the worm gear and then the 1x4 TECHNIC brick onto the axle. Make sure the axle goes through the center hole of the brick.

## Designing & Planning...

### Gear Size Matters...

This entire assembly that you are building could be made smaller. Instead of using a worm gear, attach the motor directly to the gear you are trying to turn. (However, frequently, the motor will not have enough torque for your purposes. In such cases, a worm gear is an ideal way of increasing a motor's torque.)

### Stand Motor Step 2

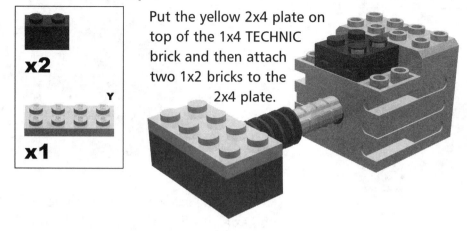

Put the yellow 2x4 plate on top of the 1x4 TECHNIC brick and then attach two 1x2 bricks to the 2x4 plate.

x2

x1
Y

### Stand Motor Step 3

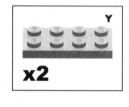

x2
Y

Turn your model upside down and attach the two yellow 2x4 plates as shown.

### Stand Motor Step 4

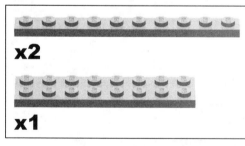

x2

x1

Attach all the plates to the bottom of the assembly as shown.

## Stand Motor Step 5

**x2**

**x1**

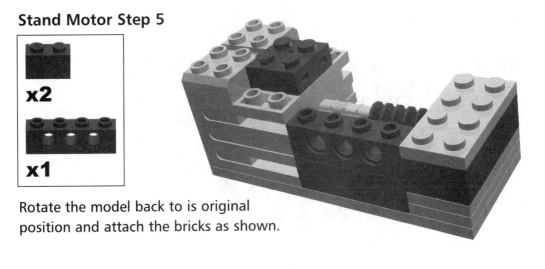

Rotate the model back to is original position and attach the bricks as shown.

## Stand Motor Step 6

**x4**

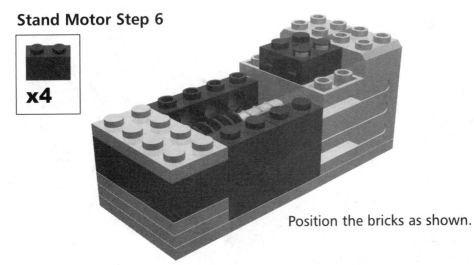

Position the bricks as shown.

## Stand Motor Step 7

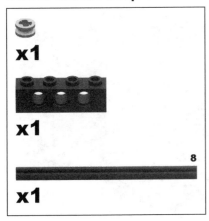

**x1**

**x1**

8

**x1**

First, you should attach a half-bushing to the axle and then slide the axle through the 1x3 TECHNIC brick. Make sure that the end of the axle is flush with the side of the 1x4 brick. Then attach the brick and axle assembly as shown.

## Stand Motor Step 8

x2

x1

Slide the 16t gear onto the axle. Make sure this gear meshes with the worm gear below it.

## Stand Motor Step 9

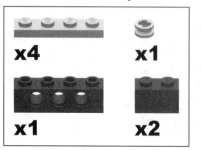

x4

x1

x1

x2

Position the bricks and plates as shown. Slide the 1x4 TECHNIC brick onto the axle. You will need to lift up the axle a little bit to allow enough room to slide the 1x4 brick over the studs and then secure it.

## Stand Motor 10

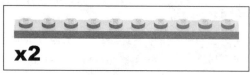

x2

Lock the entire assembly with the long 1x10 plates at the top.

# The Turntable

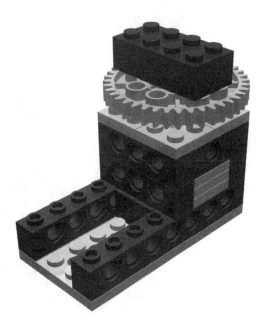

This is the turntable on which the turret assembly will sit. This turntable is powered by the motor built in the stand motor assembly.

### Turntable Step 0

x2

x1

Align the plates as shown.

### Turntable Step 1

6

x1

x1

x1

x1

First connect the TECHNIC bricks. Then, attach the bushing onto the axle and then slide the axle vertically through the hole in the plate.

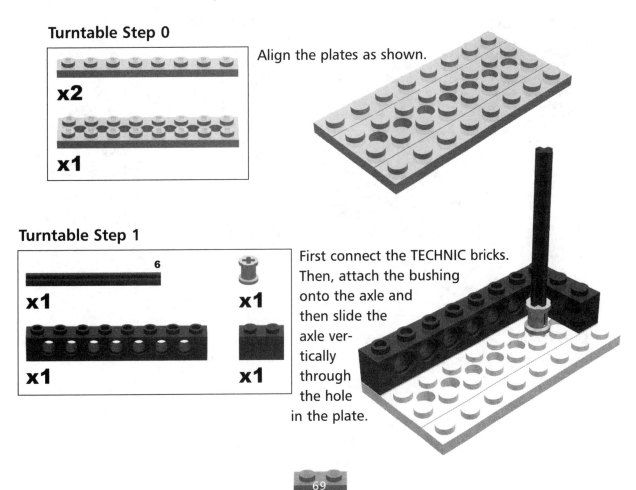

## Turntable Step 2

**x1**

**x1**

Attach the bricks as shown.

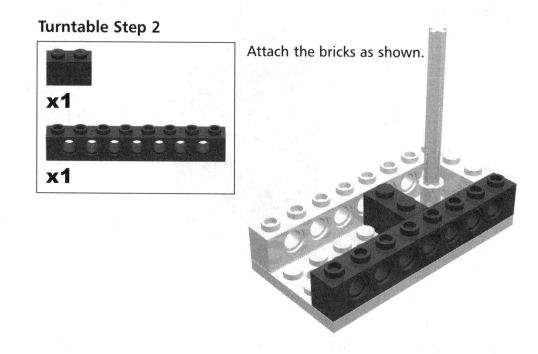

## Turntable Step 3

**x2**

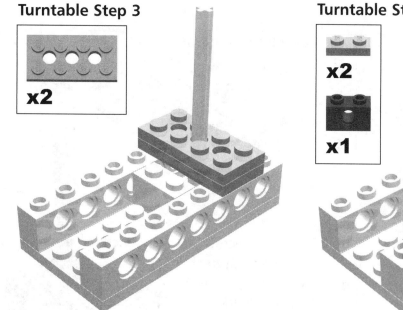

Slide the two 2x4 plates with holes onto the axle and then secure them to the bricks underneath.

## Turntable Step 4

**x2**

**x1**

Slide the 1x2 TECHNIC brick horizontally onto the axle.

## Turntable Step 5

**x2**

**x1**

**x1**

Attach the bricks as shown. The two rounded plates should be fed onto the axle with their studs facing down as shown.

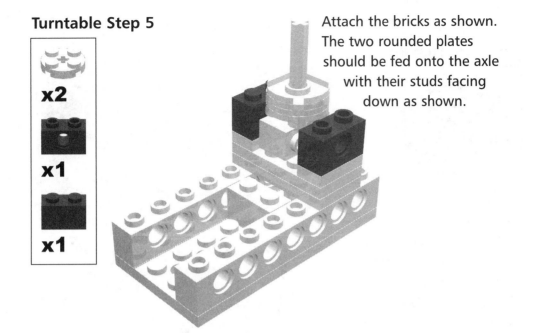

## Turntable Step 6

**x4**

Position the bricks as shown.

## Turntable Step 7

**x4**

Attach the plates as shown.

## Turntable Step 8

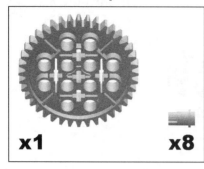

**x1**  **x8**

Slide the 40t gear onto the axle and then insert the half-pins into the gear. The pins should make a 2x4 rectangular pattern.

## Turntable Step 9

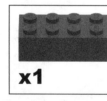

**x1**

Attach the 2x4 brick onto the half-pins on the 40t gear.

# The Stand

Now that you have constructed the motor and the turntable, you can attach the bracer sub-assemblies and build the full stand.

### Stand Step 0

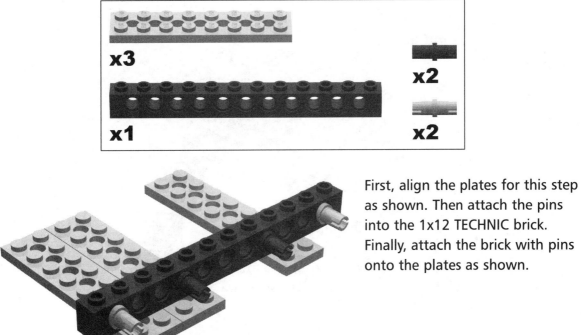

First, align the plates for this step as shown. Then attach the pins into the 1x12 TECHNIC brick. Finally, attach the brick with pins onto the plates as shown.

## Stand Step 1

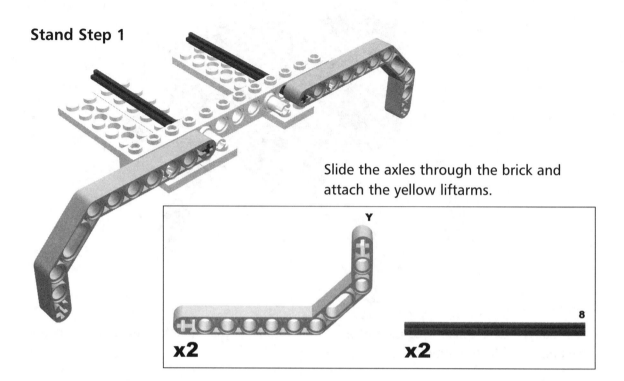

Slide the axles through the brick and attach the yellow liftarms.

## Stand Step 2

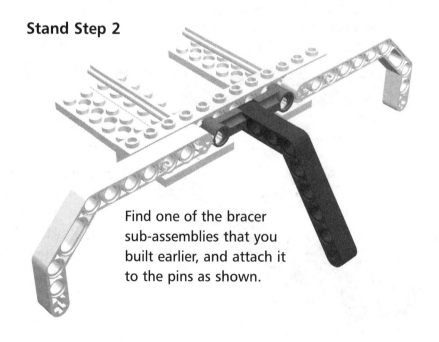

Find one of the bracer sub-assemblies that you built earlier, and attach it to the pins as shown.

## Stand Step 3

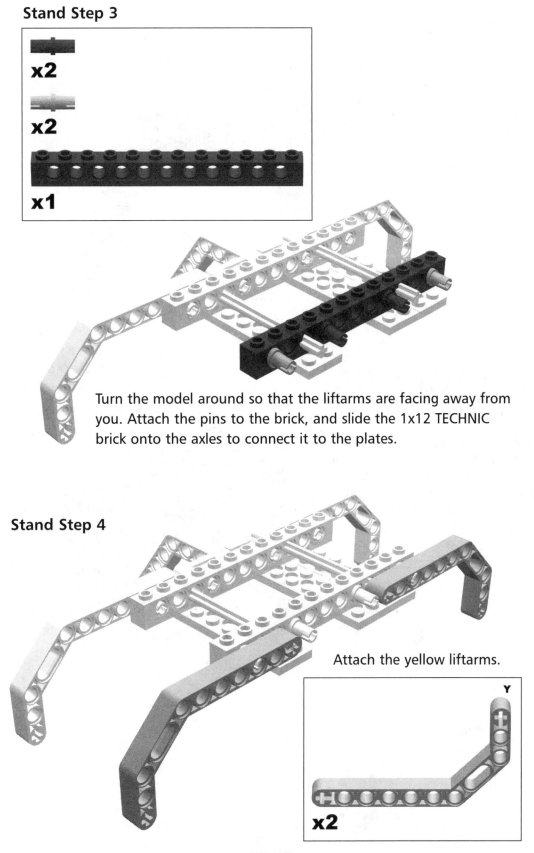

Turn the model around so that the liftarms are facing away from you. Attach the pins to the brick, and slide the 1x12 TECHNIC brick onto the axles to connect it to the plates.

## Stand Step 4

Attach the yellow liftarms.

## Stand Step 5

Find the second bracer assembly, and attach it opposite the first bracer.

Bricks & Chips...

## Bracing Considerations

Without the side bracers, the firing mechanism sitting on top of the stand may be too heavy or wide. It could cause the stand to topple over. It is always good to have a wide base as a support for higher structures sitting on top of it.

## Stand Step 6

Find the stand motor assembly, and place it on top of the stand as shown.

## Stand Step 7

**x1**

Slide a crown gear onto the protruding axle of the stand motor assembly.

## Stand Step 8

Find the turntable assembly. Holding the stand sub-assembly, slide the end of the axle holding the crown gear into the turntable assembly. Sit the turntable firmly on the stand. Finally, push the crown gear forward so that it meshes with the 40t gear.

# Cocking Mechanism

Now you are going to start building the firing mechanism–this is the part that will actually fire projectiles! We start with the assembly that will cock back the hammer.

### Cocking Mechanism Step 0

x1

x2

8

x1

x3

Attach all the parts to the axle as shown.

# Firing Hammer

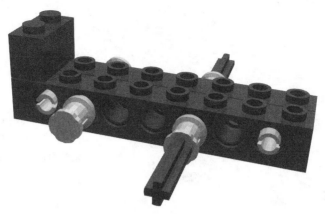

The next part of the firing mechanism you will build is the hammer. The firing hammer is what will hit your projectiles to make them fly.

### Firing Hammer Step 0

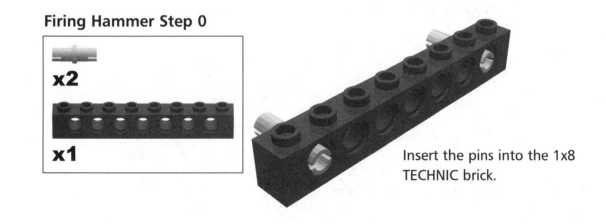

**x2**

**x1**

Insert the pins into the 1x8 TECHNIC brick.

### Firing Hammer Step 1

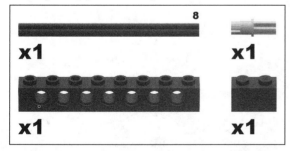

**x1**    8

**x1**

**x1**

**x1**

Attach the bricks as shown, and then insert the pins. Finally, slide the axle through the bricks.

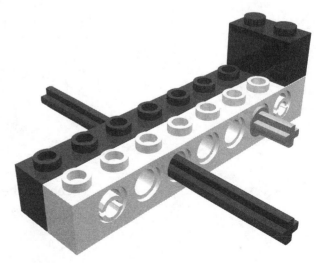

### Firing Hammer Step 2

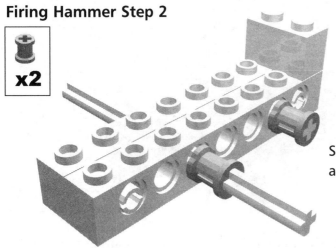

Slide the bushings onto the axle and the axle pin.

### Firing Hammer Step 3

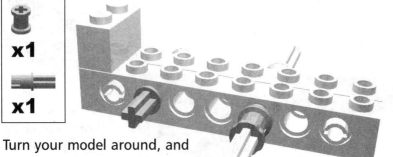

Turn your model around, and attach the bushing and the axle pin as shown.

### Firing Hammer Step 4

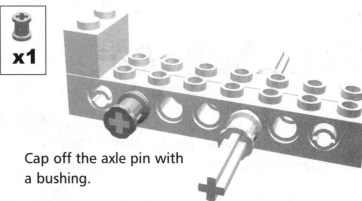

Cap off the axle pin with a bushing.

# The Firing Motor

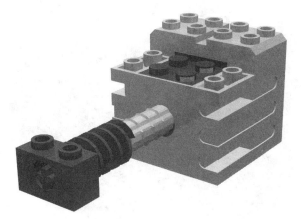

This is the motor assembly that will power the firing of your projectiles. It is braced sturdily so that the assembly will not come apart if it meets resistance.

### Firing Motor Step 0

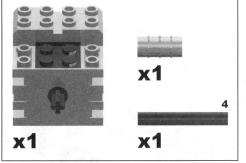

x1

x1

4

x1

Attach the axle joiner to the motor, and then attach the axle to the axle joiner.

### Firing Motor Step 1

x1

x1

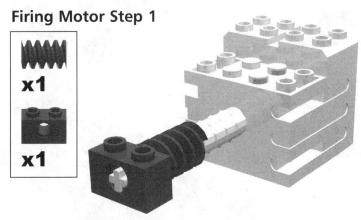

Slide the worm gear and then the 1x2 brick through the axle.

# Firing Motor Casing

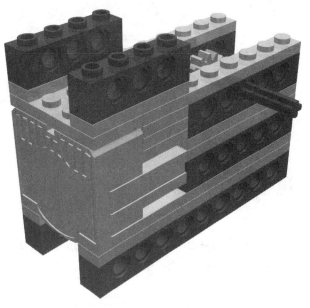

Now that you have the firing motor, you will build the casing assembly around it.

## Firing Motor Casing Step 0

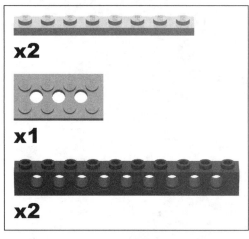

**x2**

**x1**

**x2**

Construct the base of the motor assembly by connecting the bricks and plates as shown.

## Firing Motor Casing Step 1

**G**

**x1**

Attach the green 2x2 plate as shown.

## Firing Motor Casing Step 2

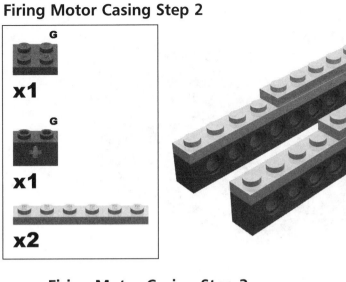

**x1** G

**x1** G

**x2**

Place another green 2x2 plate on top of the existing green plate added in **Firing Motor Casing Step 1**. Attach the rest of the parts as shown.

## Firing Motor Casing Step 3

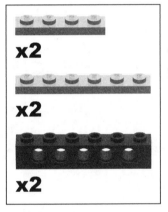

**x2**

**x2**

**x2**

Attach the bricks and plates.

## Firing Motor Casing Step 4

**x1**

Find the firing motor assembly and place it directly on top of your model and secure firmly. Make sure the wire of the wire connector brick is facing out the backside of the motor. The wire should sit in the groove at the top of the motor.

## Firing Motor Casing Step 5

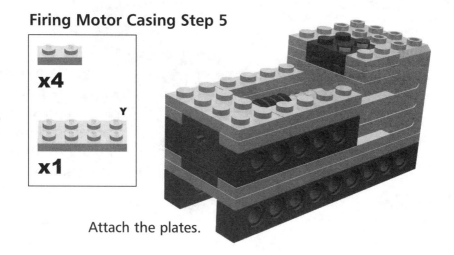

**x4**

Y

**x1**

Attach the plates.

## Firing Motor Casing Step 6

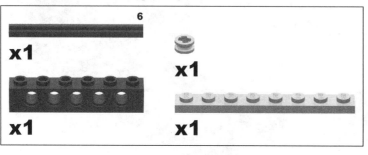

6

**x1**

**x1**

**x1**

**x1**

Attach the bricks and plates. Slide the
half-bushing onto the axle, and then
slide the axle through the
middle hole of the 1x6 brick.

### Firing Motor Casing Step 7

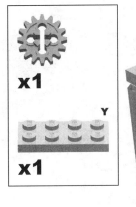

**x1**

**x1**

Slide the 16t gear onto the axle until it hits the half-bushing. Attach the yellow 2x4 plate as shown.

### Firing Motor Casing Step 8

**x1**

Slide a half-bushing onto the axle until it meets the 16t gear.

### Firing Motor Casing Step 9

Slide the brick through onto the axle. You will then have to lift up the axle a bit so that the 1x6 TECHNIC brick can be attached to the studs underneath the axle.

**x1**

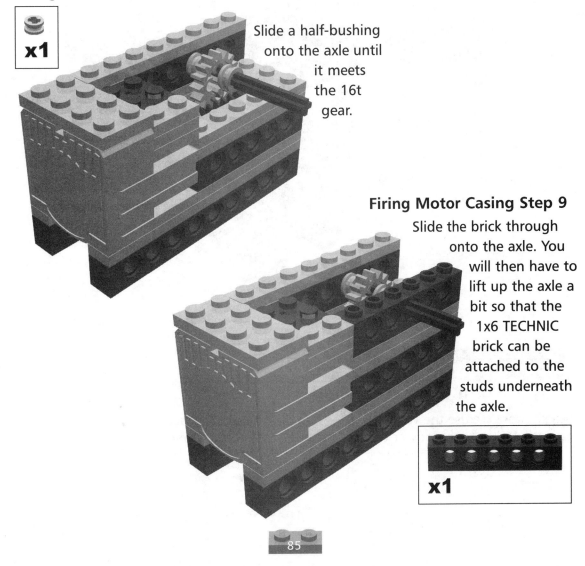

## Firing Motor Casing Step 10

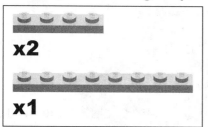

**x2**

**x1**

Attach the plates as shown.

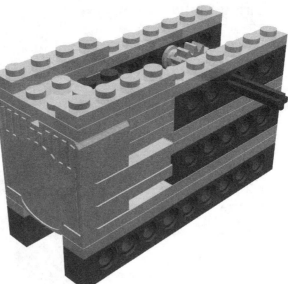

## Firing Motor Casing Step 11

**x2**

Attach the two 1x4 TECHNIC bricks as shown. The firing motor assembly is complete.

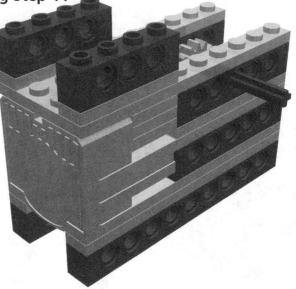

### Design & Planning...

## Why a Worm Gear?

Again, here we use the worm gear for lots of torque. You'll see that this is just barely enough power to cock the hammer and fire projectiles!

# The Right Side of the Turret

Now you will start to build the frame of the firing mechanism. This will hold all the motors and firing parts steady. You will start with the right side of the frame.

### Right Side of Turret Step 0

x4

x1

Insert the long pins with friction into the 1x10 TECHNIC brick as shown.

## Right Side of Turret Step 1

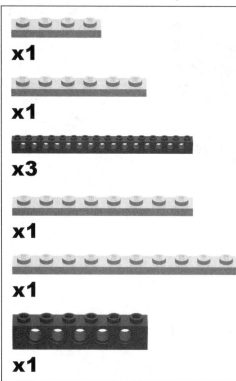

**x1**

**x1**

**x3**

**x1**

**x1**

**x1**

Connect all the bricks and plates as shown. Note that you should connect the beams and plates first, and then slide them onto the long pins with friction.

## Right Side of Turret Step 2

**x2**

**x2**

Insert the axle pins and pins as shown.

## Right Side of Turret Step 3

**x2**

**x1**

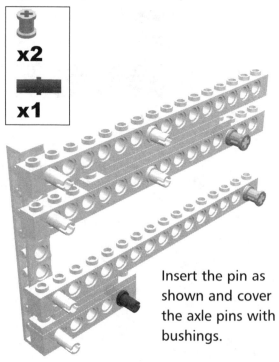

Insert the pin as shown and cover the axle pins with bushings.

# The Full Turret

Now you will attach the motor to the right side of the frame and then construct the rest of the firing mechanism around it.

## Full Turret Step 0

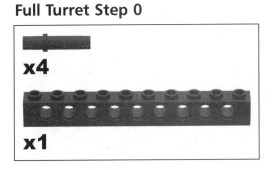

x4

x1

Insert the pins into the brick as shown.

## Full Turret Step 1

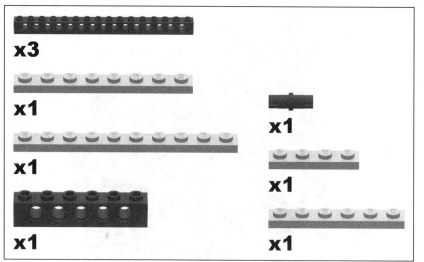

x3

x1

x1

x1

x1

x1

x1

Attach the bricks and plates as shown. Attach the TECHNIC bricks to the plates before you connect the bricks to the long pins.

## Full Turret Step 2

x2

x2

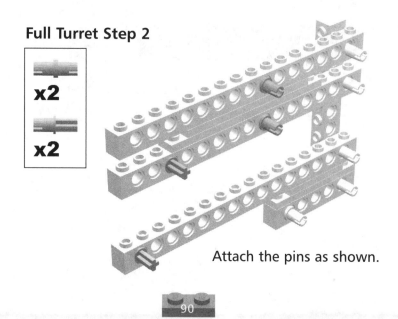

Attach the pins as shown.

## Full Turret Step 3

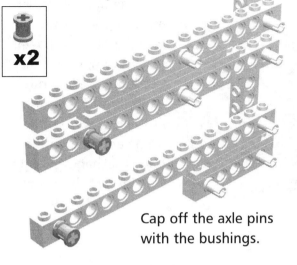

Cap off the axle pins with the bushings.

## Full Turret Step 4

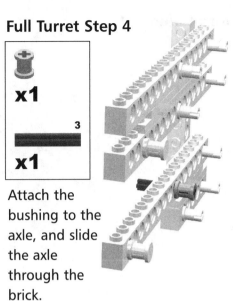

Attach the bushing to the axle, and slide the axle through the brick.

## Full Turret Step 5

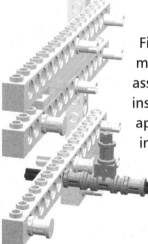

Find the cocking mechanism assembly, and insert it into the appropriate hole in the brick.

## Full Turret Step 6

Slide the gears onto the axles.

## Full Turret Step 7

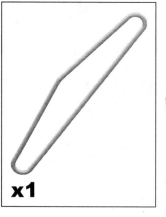

**x1**

Find the firing hammer assembly, and attach it to the model as shown. It is very important that you loop a yellow rubber band through all three bushings (two on the frame and one on the firing mechanism) on the left side of this assembly.

## Full Turret Step 8

**x1**

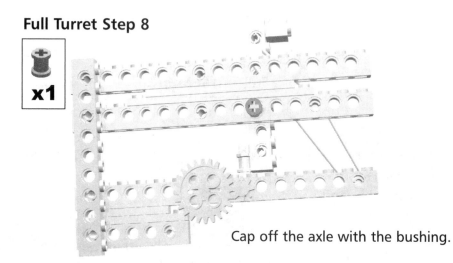

Cap off the axle with the bushing.

## Full Turret Step 9

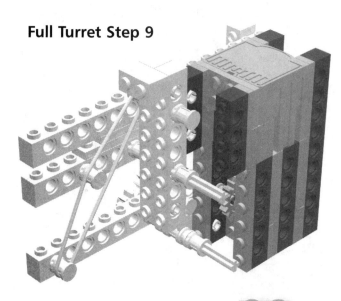

Find the firing motor casing assembly, and attach it vertically as shown. Make sure the axle slides through the frame.

## Full Turret Step 10

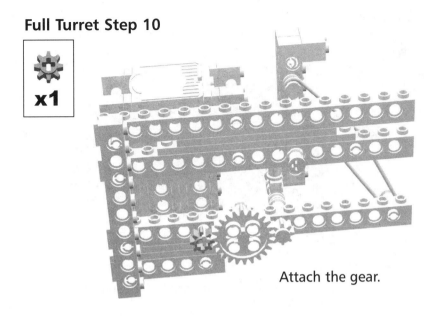

**x1**

Attach the gear.

## Full Turret Step 11

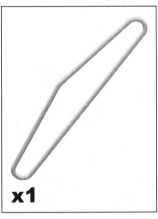

**x1**

Take the right side of the assembly and attach it as shown. It is very important that you loop another yellow rubber band through the three bushings on the right side of the frame. The result should look the same as in **Full Turret Step 7**.

## Full Turret Step 12

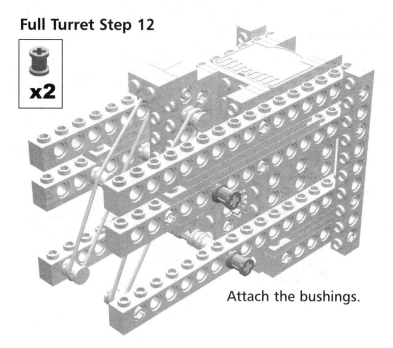

Attach the bushings.

## Full Turret Step 13

Attach the plates as shown.

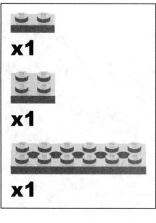

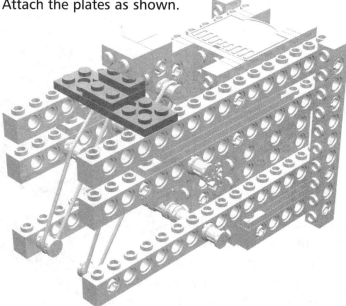

## Full Turret Step 14

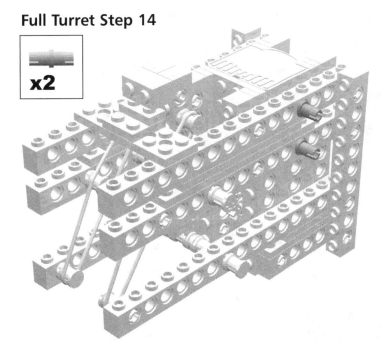

**x2**

Attach the pins as shown.

## Full Turret Step 15

**x3**

10

**x1**

**x1**

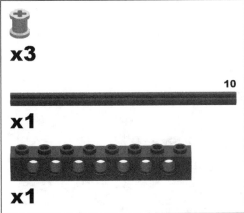

Attach the bricks, axle, and bushings as shown.

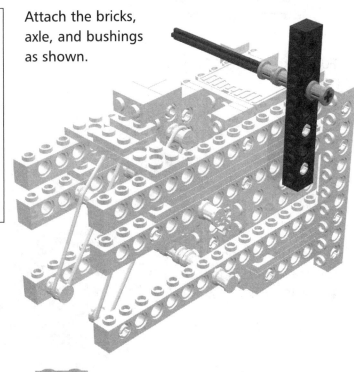

## Full Turret Step 16

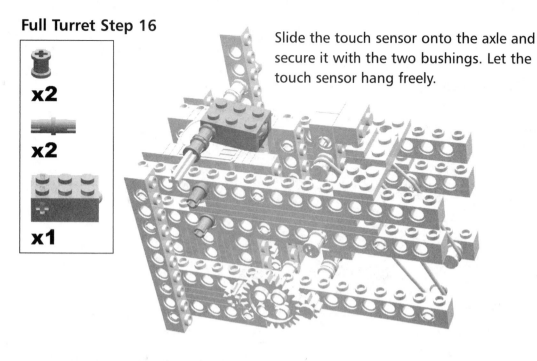

Slide the touch sensor onto the axle and secure it with the two bushings. Let the touch sensor hang freely.

Inventing...

## Touch Sensor Placement

The touch sensor here will be pressed when the hammer is fully cocked, just before it is about to fire. You can use this in your programs to pre-wind the hammer so that your turret will be prepared for a target and can fire right away!

## Full Turret Step 17

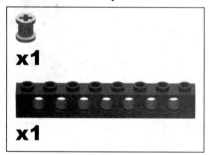

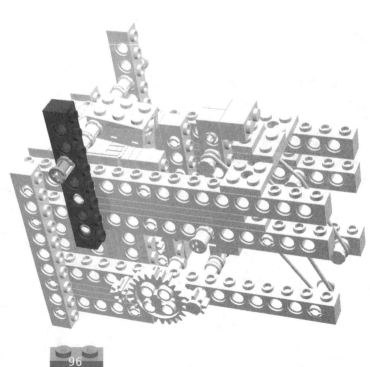

Secure the axle on which the touch sensor hangs with the TECHNIC brick and a bushing.

## Full Turret Step 18

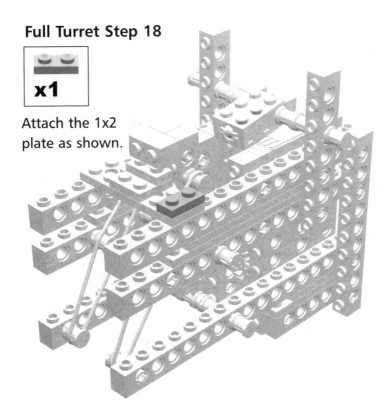

**x1**

Attach the 1x2
plate as shown.

## Full Turret Step 19

**x1**

**x2**

**x1**

8

**x1**

**x1**

**x2**

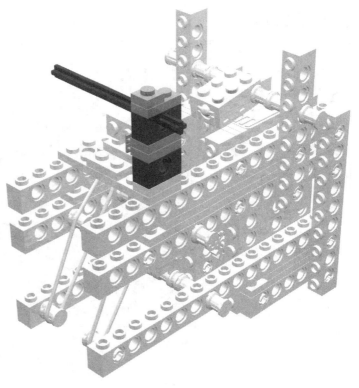

Stack the bricks and
plates, and slide the
axle through.

## Full Turret Step 20

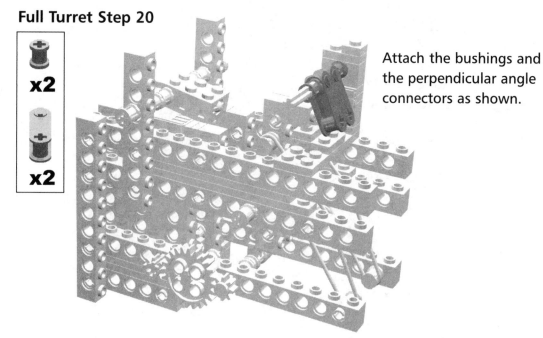

Attach the bushings and the perpendicular angle connectors as shown.

## Full Turret Step 21

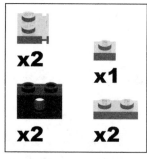

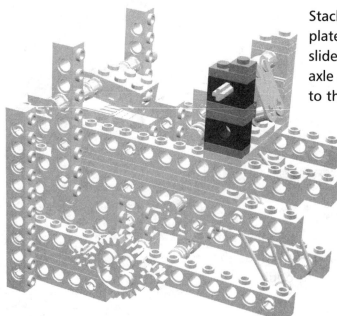

Stack the bricks and plate together, then slide them onto the axle and secure them to the frame.

## Full Turret Step 22

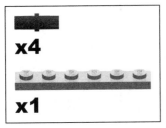

**x4**

**x1**

Insert the pins and attach the plate as shown.

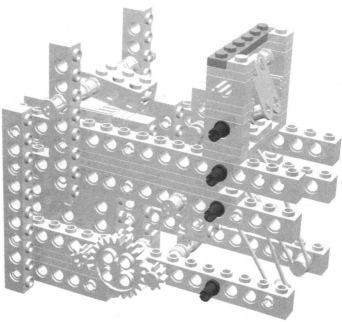

## Full Turret Step 23

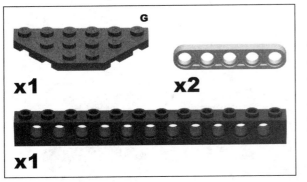

**x1** G

**x2**

**x1**

Attach bricks and plates as shown. The beams in this step provide a fairly smooth launching strip for your projectiles.

## Full Turret Step 24

**x3**

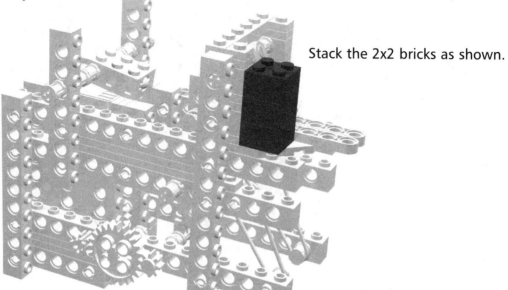

Stack the 2x2 bricks as shown.

## Full Turret Step 25

**x2**

**x1**

**x1**

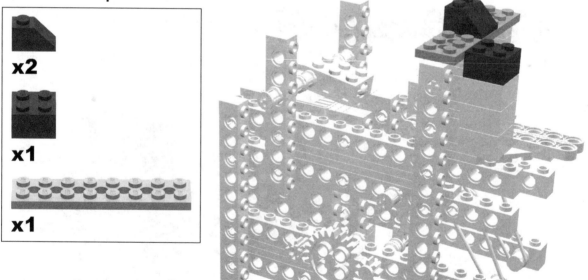

Stack the bricks and plates as shown.

## Full Turret Step 26

x4

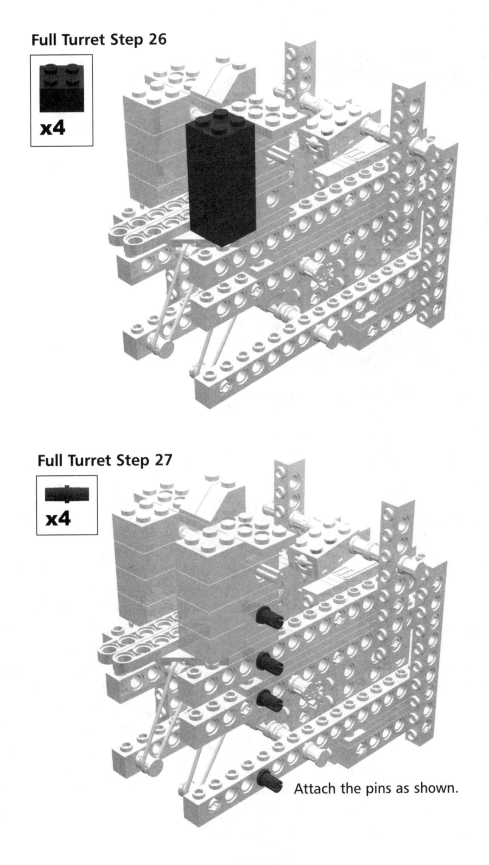

## Full Turret Step 27

x4

Attach the pins as shown.

## Full Turret Step 28

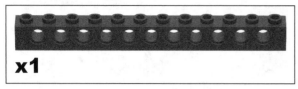

**x1**

Attach the brick to the pins.

## Full Turret Step 29

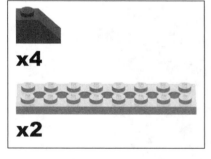

**x4**

**x2**

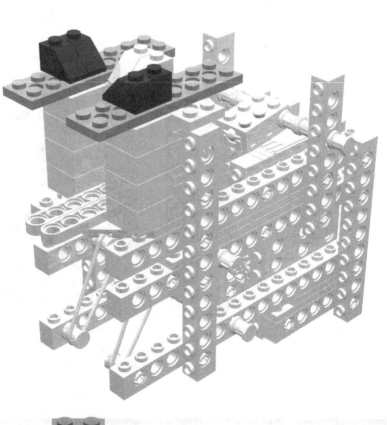

## Full Turret Step 30

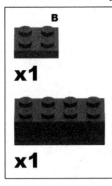

**x1**

**x1**

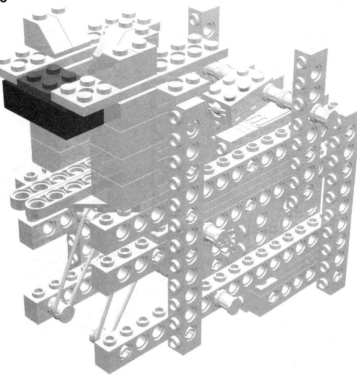

Attach the 2x4 brick first and then place the blue 1x2 plate on top.

## Full Turret Step 31

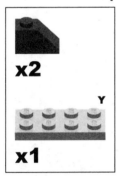

**x2**

**x1**

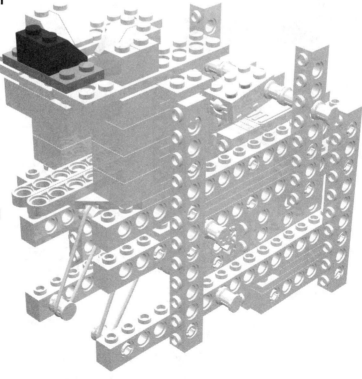

Attach the yellow 2x4 plate and sloped bricks as shown.

## Full Turret Step 32

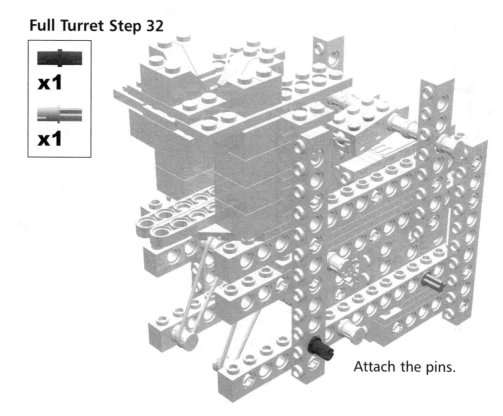

x1

x1

Attach the pins.

## Full Turret Step 33

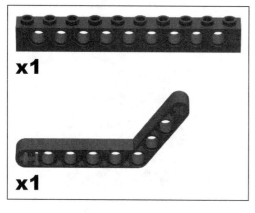

x1

x1

Attach the liftarm and the brick. Let them hang loose for now.

### Full Turret Step 34

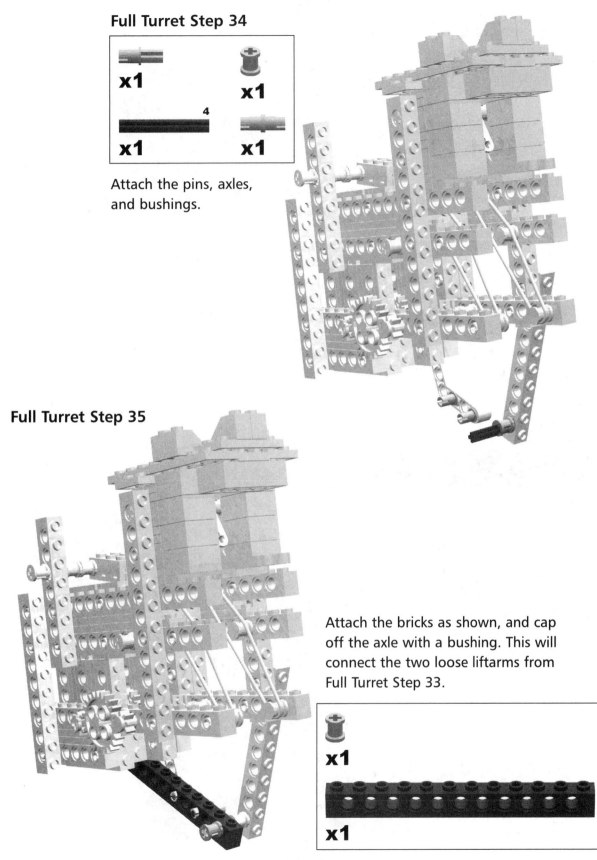

Attach the pins, axles, and bushings.

### Full Turret Step 35

Attach the bricks as shown, and cap off the axle with a bushing. This will connect the two loose liftarms from Full Turret Step 33.

## Full Turret Step 36

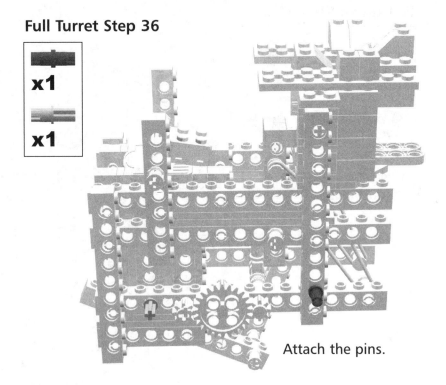

x1

x1

Attach the pins.

## Full Turret Step 37

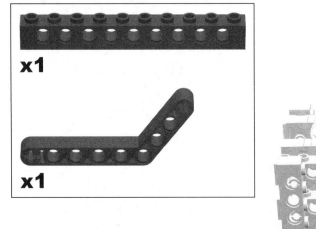

x1

x1

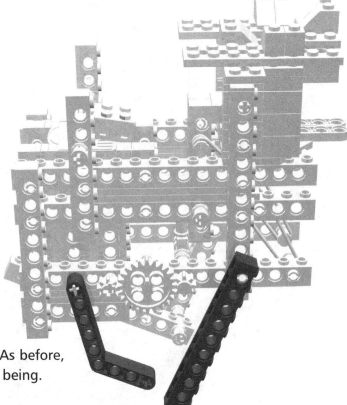

Attach the liftarm and the brick. As before, let these hang loose for the time being.

## Bricks & Chips...

## Liftarms and Angles

Using various types of liftarms, you can get a variety of angles into your creations. Here, we want to elevate the turret a little bit so we use a liftarm to create a non-perpendicular angle.

**Full Turret Step 38**

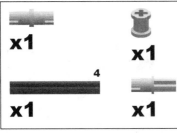

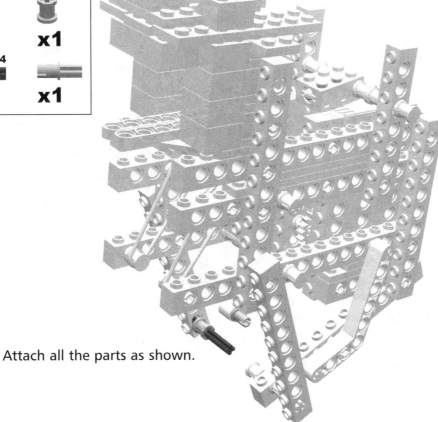

Attach all the parts as shown.

## Full Turret Step 39

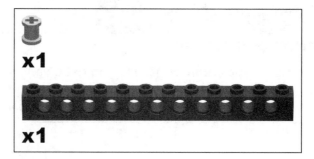

Connect the liftarm and the brick from Full Turret Step 37 using the 1x12 TECHNIC brick.

## Full Turret 40

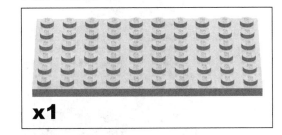

**x1**

Attach the 6x10 plate to the underside of the bricks.

# Final Assembly

Finally, now that you have built both the firing mechanism and the stand, you can put them together.

Find the stand you built. Mount the firing mechanism on the stand, and you're all done!

Inventing...

## Programming Ideas

Now that you have a working, rotating missile turret, test it out: Put a small projectile into the top hole and watch it fire! Can you design a program to cock the hammer fully and then wait and fire? How about mounting a light sensor on the turret so that it will search for light and then fire a projectile when it finds a strong enough light source?

# Robot 4

## MINDSTORMS F1 Racer

Although a car is a very common type of vehicle, it is built very much unlike a common robotic vehicle. Robotic vehicles are usually built so that separate motors control the wheels on each side. This setup is advantageous because it is easy to build, and navigation is easily controlled through software that has direct control over each wheel. Going forward, turning, and going in reverse can all be achieved by a simple combination of forward/reverse commands to the motors. A car, on the other hand, is usually not built this way. The motors are not attached directly to the wheels; rather, one motor is devoted to driving the car forward. This means that the motor must be able to steer the entire car without access to the driving wheels.

This MINDSTORMS model is a simplification of a common car. Like a car, this model is driven by its rear wheels through the use of a *differential*. It is steered by a *rack and pinion* system, also like a real car. One motor is used to drive the back wheels and one to steer the car. The chassis gives the model stability and body work is added for the final decorative touch.

This car is built in stages, also somewhat like a real car. You will first build the chassis, then add individual components (driving motor, rack and pinion steering system, steering motor, front bumper, and a spoiler) to the chassis, and then put the wheels on to complete the car. The RIS kit does not include many steering or driving parts available in other TECHNIC sets, so you will have to be creative with the parts you do have.

After you finish building your car, you can make some simple programs to drive it around. For an extra challenge, try to add a bumper system so that your car can react to obstacles!

**NOTE**

For a more detailed model of a car, check out LEGO TECHNIC Model #8448: The Super Street Sensation.

# The Right Chassis

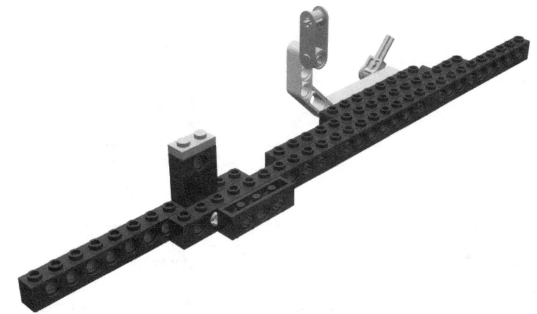

The first part you will build is the right side of the chassis. When combined with the left side of the chassis, this will form the structural base of the car.

### Right Chassis Step 0

**x1**

This part is called a pole reverser handle. In future steps, you will have to connect flexible elements to this part.

### Right Chassis Step 1

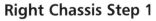

**x1**

Insert the axle pin into the pole reverser handle.

## Bricks & Chips...
### Pins & Friction

When building with the long pin with friction, as in this step, make sure that the shorter segmented end goes into the element that you are first attaching it to. Otherwise, the long pins will slide through.

## Right Chassis Step 2

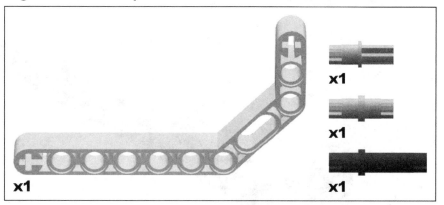

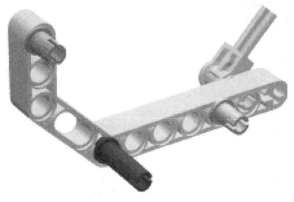

Attach the yellow lift arm and pins to the pole reverser handle. Let the pole reverser handle dangle.

## Right Chassis Step 3

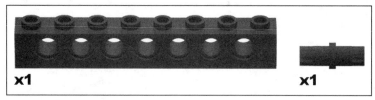

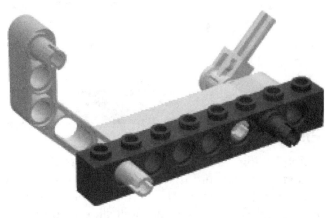

Connect the TECHNIC brick and pins to the liftarm. Make sure the long pin from the previous step comes through the brick.

## Right Chassis Step 4

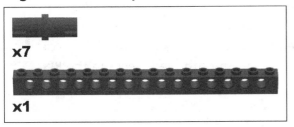

**x7**

**x1**

Connect another brick to what you have built already. Make sure that you insert all the pins as shown.

## Right Chassis Step 5

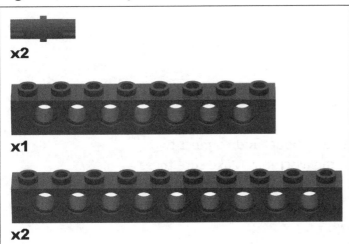

**x2**

**x1**

**x2**

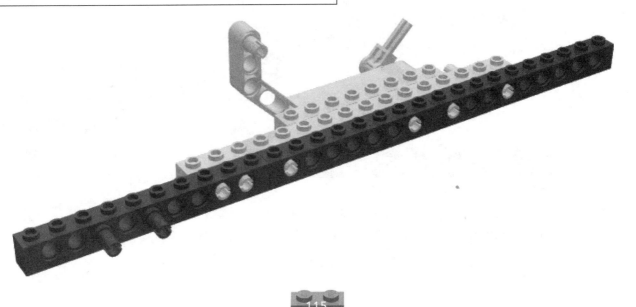

## Right Chassis Step 6

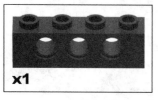

**x1**

Connect the 1x4 brick. Make sure the brick is upside down in relation to the rest of the assembly. The 1x4 brick is mounted upside down for a reason: Since there are no smooth-tiled pieces in the RIS kit, you can use the underside of bricks as a substitute.

## Right Chassis Step 7

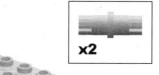

**x2**

Rotate your model and connect two pins to the assembly. Make sure these are normal pins, not pins with friction.

### Bricks & Chips...

## Pins & Rotation

If you want a part to rotate freely, use the normal pins. Otherwise, use the pins with friction to get extra gripping power.

### Right Chassis Step 8

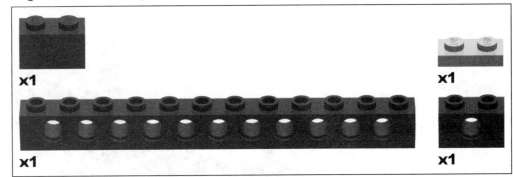

x1  x1  x1  x1

### Right Chassis Step 9

x1

Connect the perpendicular angle connector to the assembly to finish the left side of the chassis.

# The Left Chassis

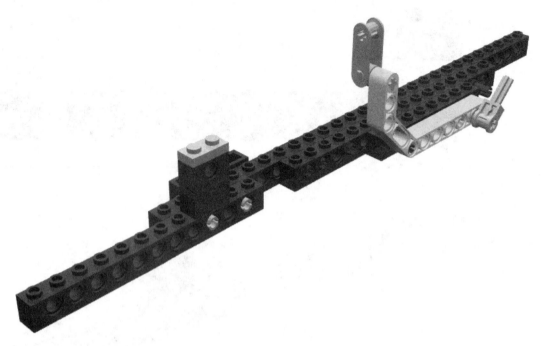

This is what the left side of the chassis will look like when finished. It should be a mirror image of the right side of the chassis.

### Left Chassis Step 0

x1                    x2

Start with the 1x4 brick.
Make sure this brick remains
upside down in relation to
the rest of the assembly.

## Left Chassis Step 1

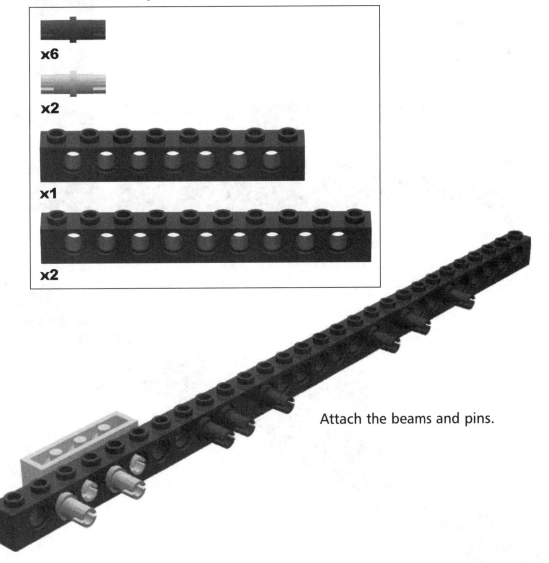

x6

x2

x1

x2

Attach the beams and pins.

## Left Chassis Step 2

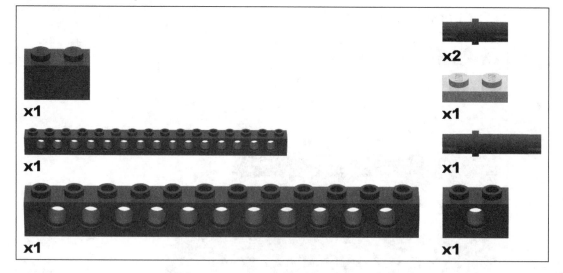

Attach more beams and pins.

## Left Chassis Step 3

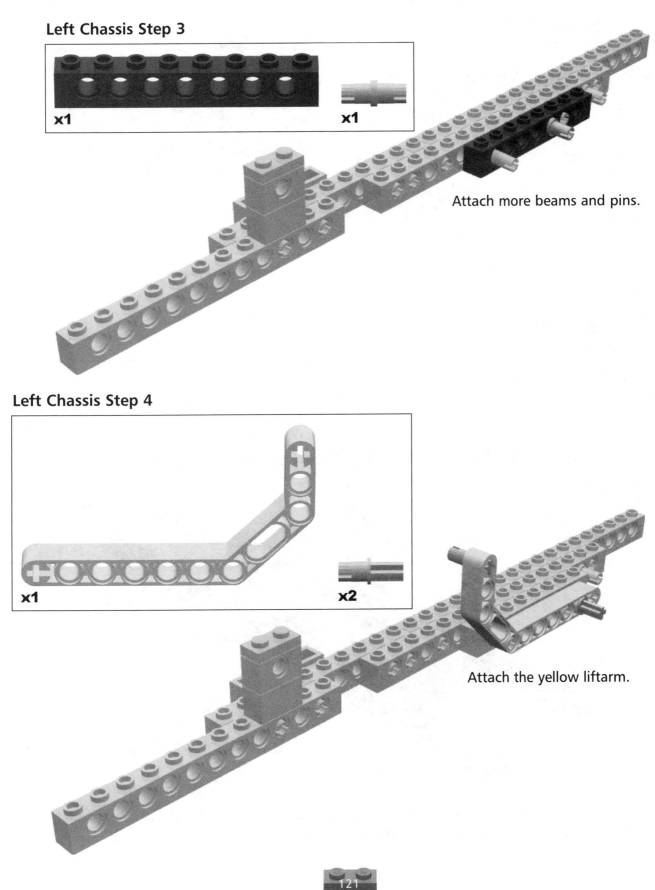

x1  x1

Attach more beams and pins.

## Left Chassis Step 4

x1  x2

Attach the yellow liftarm.

## Left Chassis Step 5

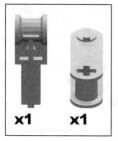

x1     x1

Attach the last two parts to finish the
right side of the chassis. Let these two
parts hang—don't worry about their
positions for now.

# Connecting the Left and Right Sides of the Chassis

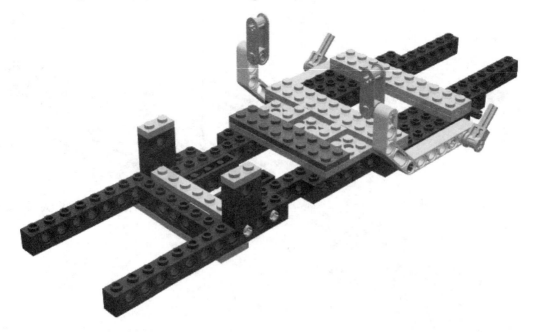

Here is where things start coming together to form the framework of the car. The next steps will
connect the left and right side of the chassis to each other.

## Connecting Chassis Step 0

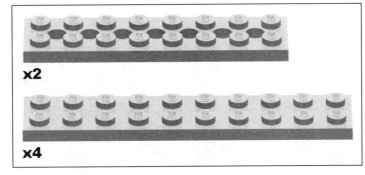

**x2**

**x4**

Find the completed right side of the chassis that you built earlier. Connect the plates to the right side of the chassis.

## Connecting Chassis Step 1

**x2**

**x1**

Add more plates. The color is not important in this step, so long as the plate is a 2x8 plate without holes.

## Connecting Chassis Step 2

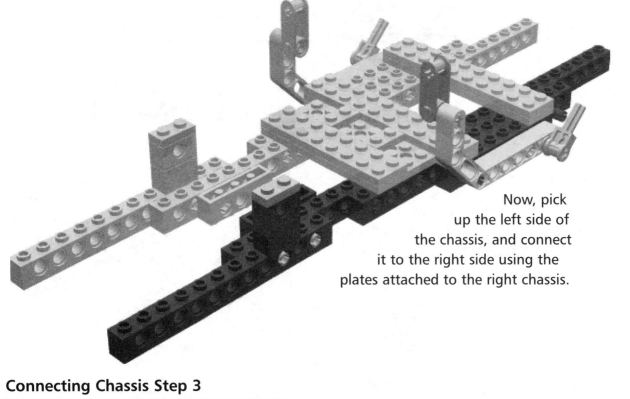

Now, pick
up the left side of
the chassis, and connect
it to the right side using the
plates attached to the right chassis.

## Connecting Chassis Step 3

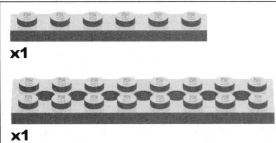

**x1**

**x1**

Stabilize the
front of the complete
chassis with beams.

## Connecting Chassis Step 4

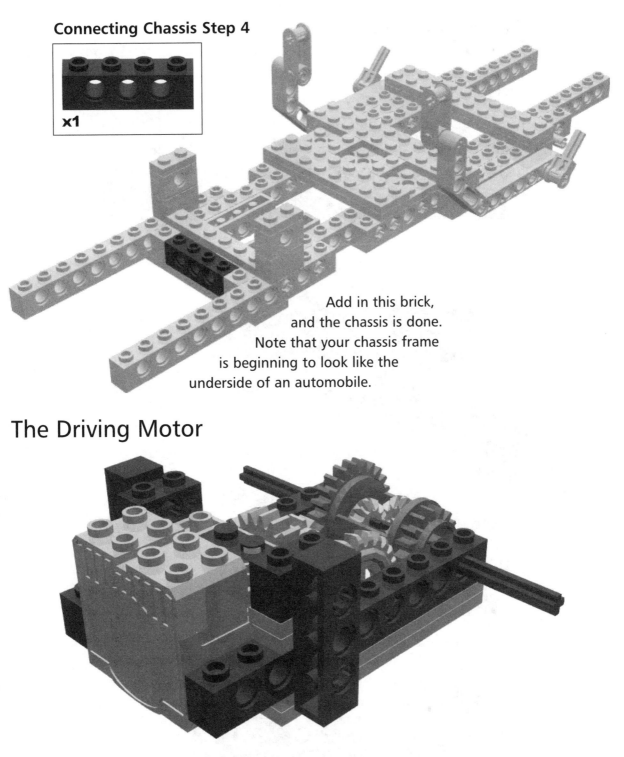

**x1**

Add in this brick,
and the chassis is done.
Note that your chassis frame
is beginning to look like the
underside of an automobile.

# The Driving Motor

This next series of steps will create the differential drive motor. This motor is what will drive the car. The assembly consists of a TECHNIC 9V motor and a small gear train used to change direction and transfer the power to the differential.

## Driving Motor Step 0

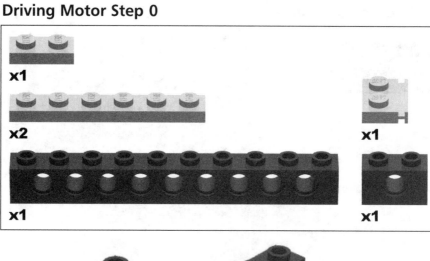

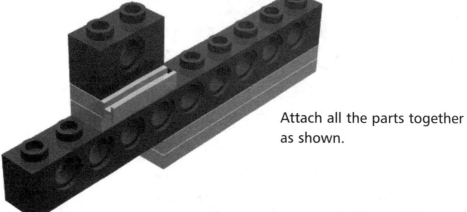

Attach all the parts together as shown.

## Driving Motor Step 1

**x2**

Turn around your model from step Driving Motor 0, and connect the pins. Make sure that the plate with door rails faces opposite the pins.

## Driving Motor Step 2

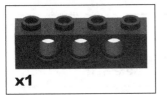

Connect the 1x4 brick to the pins; this functions as a vertical bracing beam.

## Driving Motor Step 3

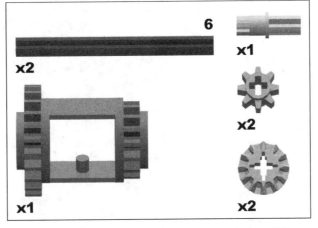

Add in part of the differential.

## Bricks & Chips...

## Securing Motors

Most assemblies built to secure motors will need vertical bracing beams so that the stress of the motor doesn't rip the assembly apart.

## Driving Motor Step 4

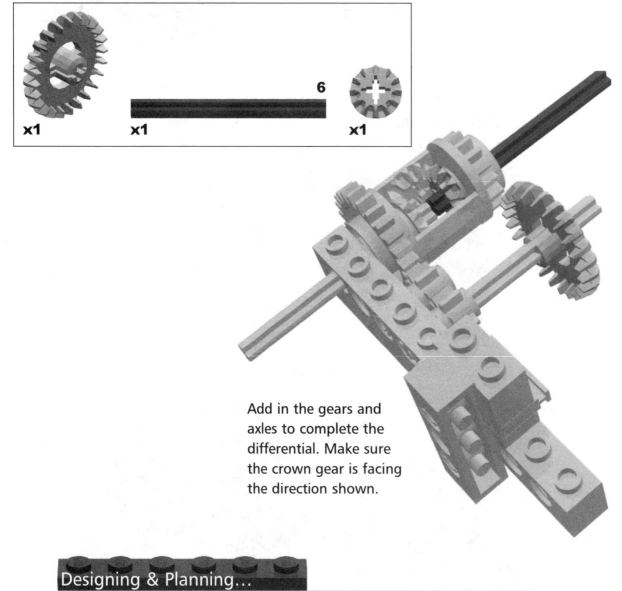

Add in the gears and axles to complete the differential. Make sure the crown gear is facing the direction shown.

## Designing & Planning...

## Driving Motor Step 3

In this step, you could substitute the differential for a straight axle with a 24t gear on one side. This would make the back wheels drive at exactly the same rates regardless of the road conditions.

## Driving Motor Step 5

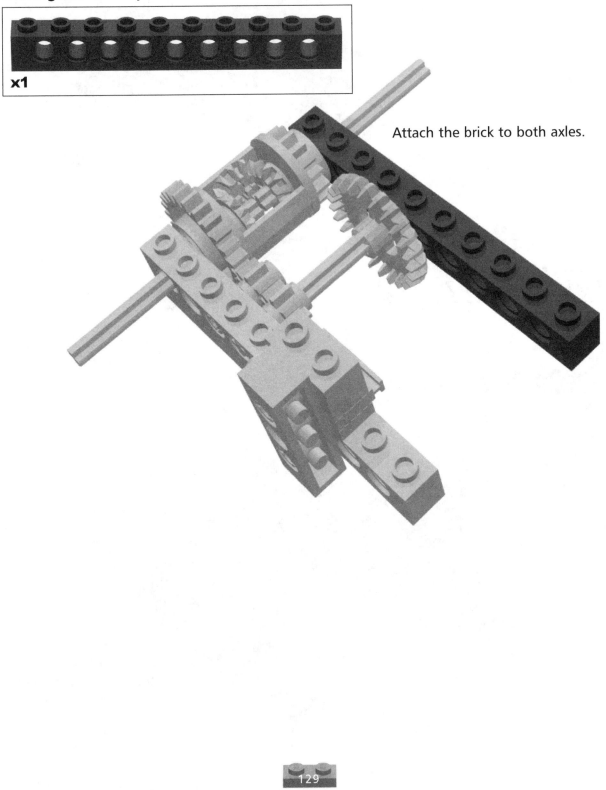

**x1**

Attach the brick to both axles.

## Driving Motor Step 6

x1   x1   x1

Attach the bricks and plates.

## Driving Motor Step 7

x2

x2

Attach the pins and
the plates underneath.

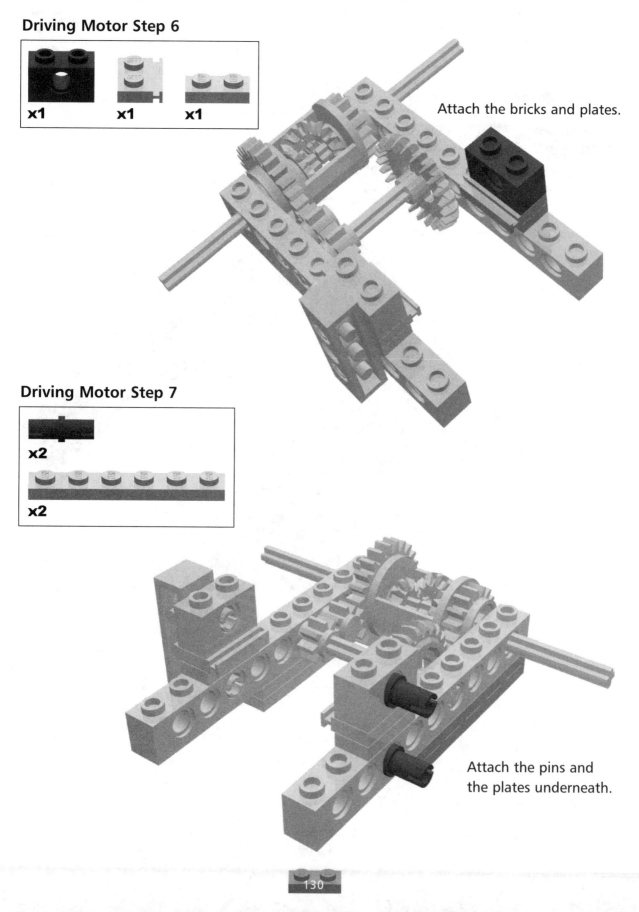

### Driving Motor Step 8

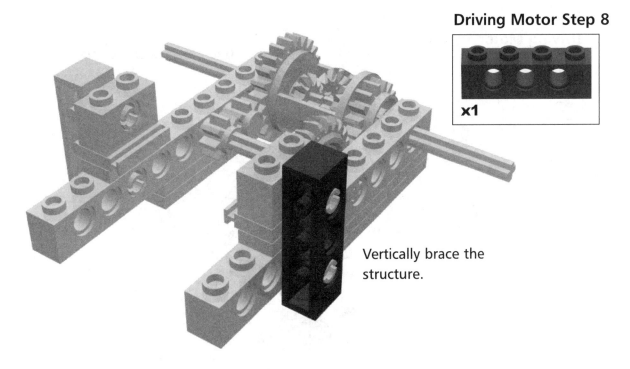

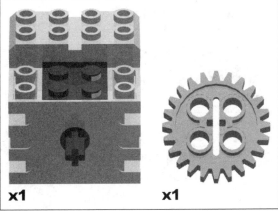

x1

Vertically brace the structure.

### Driving Motor Step 9

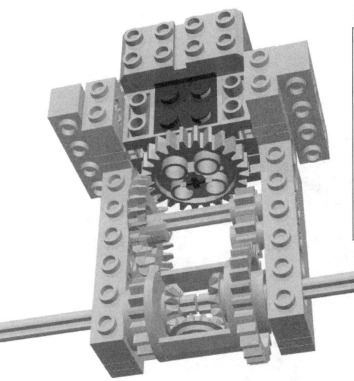

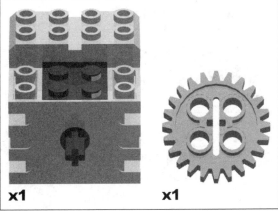

x1          x1

Attach the gear to the motor, and insert the motor into the assembly.

## Driving Motor Step 10

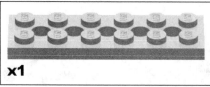

**x2**

Attach the 1x2 spacer plates.

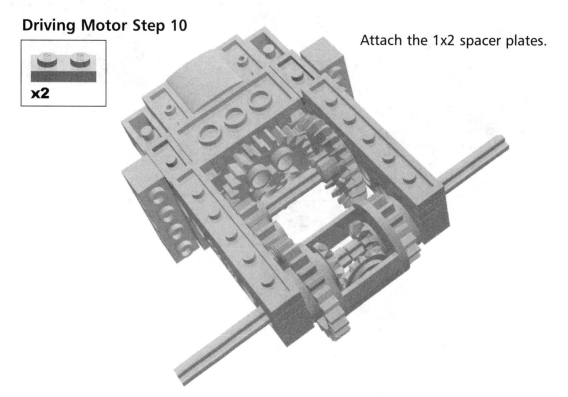

## Driving Motor Step 11

**x1**

Secure the motor and
assembly horizontally
with a 2x6 plate.

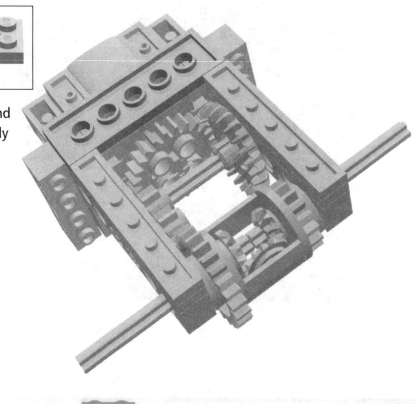

# The Steering Mechanism

This is the rack-and-pinion steering system. The rack will be free to move left and right, while the other side will be secured to the chassis. The left and right motion of the rack will direct the axles, which in turn will control the front wheels.

### Steering Mechanism Step 0

**x1**          **x2**

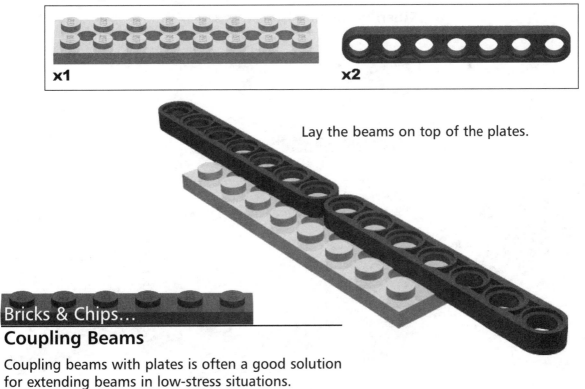

Lay the beams on top of the plates.

## Bricks & Chips...
## Coupling Beams

Coupling beams with plates is often a good solution for extending beams in low-stress situations.

## Steering Mechanism Step 1

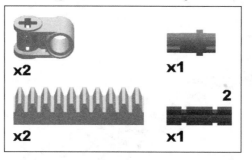

Attach the rack, and get some parts ready. This is the rack that will eventually steer the car.

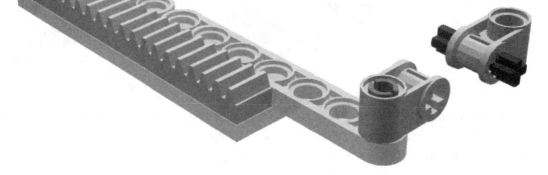

## Steering Mechanism Step 2

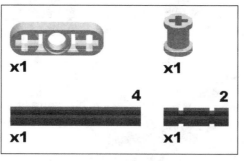

Attach the beams and axles.

## Steering Mechanism Step 3

**x1**

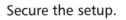

Secure the setup.

## Steering Mechanism Step 4

**x1**

Attach the half-length pin.

## Steering Mechanism Step 5

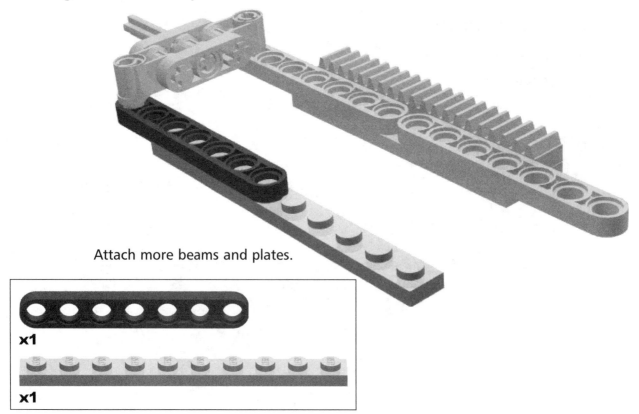

Attach more beams and plates.

x1

x1

## Steering Mechanism Step 6

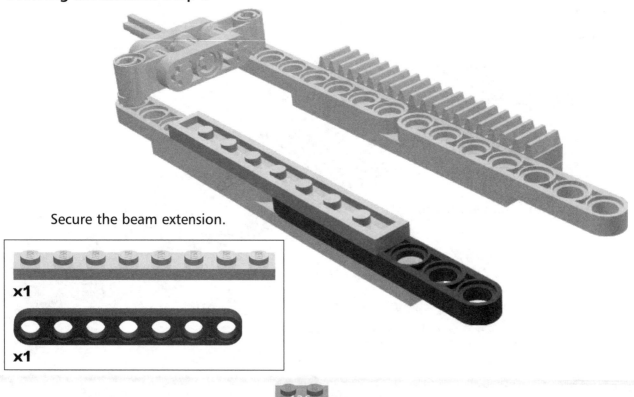

Secure the beam extension.

x1

x1

## Steering Mechanism Step 7

**x2**

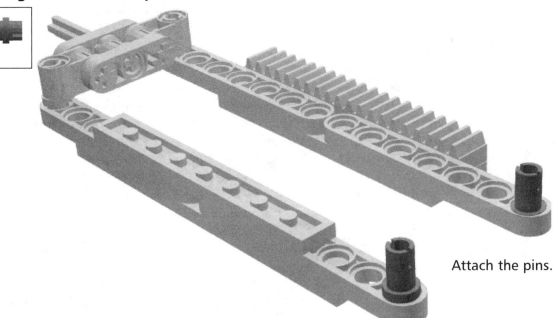

Attach the pins.

## Steering Step 8

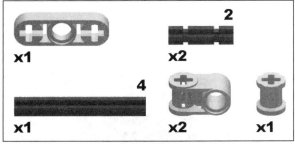

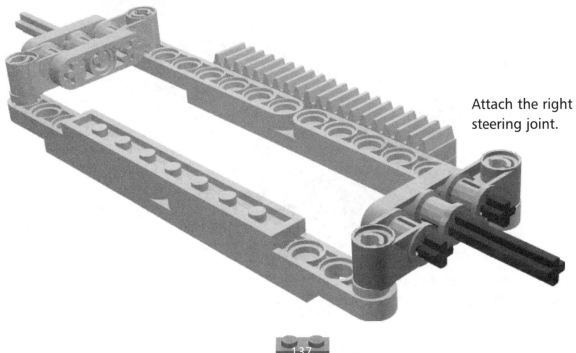

Attach the right steering joint.

## Steering Mechanism Step 9

**x1**

Complete the right steering joint and steering assembly. Take a moment to make sure that this assembly can slide left and right.

# The Steering Motor

This is the motor that will steer your car. It will be placed over the rack. The 16t gears will mesh with the rack and steer the car left and right.

## Steering Motor Step 0

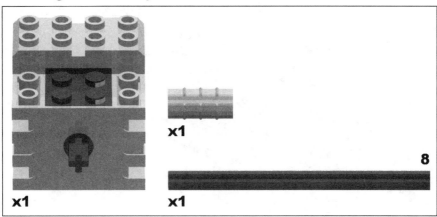

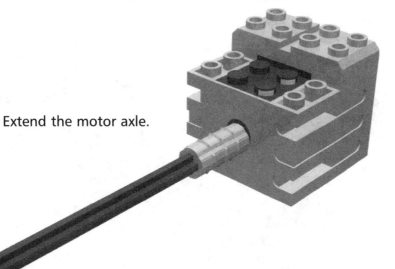

Extend the motor axle.

Bricks & Chips...

## Gears & Motors

Often, putting gears directly on the motor will not be strong enough: Torque and friction will make the gear slide off the short axle. In these cases, it is best to extend the motor axle, put the gears on the extended axle, and, if necessary, brace the axle from the other side.

## Steering Motor Step 1

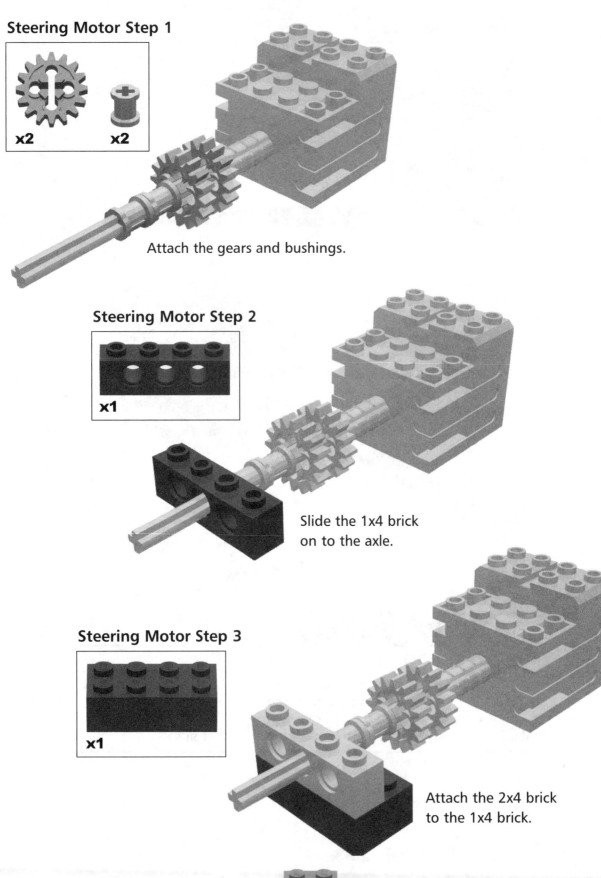

x2   x2

Attach the gears and bushings.

## Steering Motor Step 2

x1

Slide the 1x4 brick
on to the axle.

## Steering Motor Step 3

x1

Attach the 2x4 brick
to the 1x4 brick.

## Steering Motor Step 4

**x1**

Attach the plate and complete the steering motor assembly.

# The Bumper

This is the front bumper. While it is non-functional in this car, it does make the model much more convincing. Headlights included!

Inventing...

## Optimizing Your Racer

A good improvement to this car would be to add touch sensor(s) to this bumper. Can you design a functional bumper system for this robot?

## Bumper Step 0

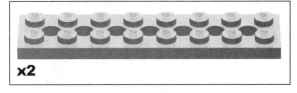

**x2**

Stack the plates on top of each other.

## Bumper Step 1

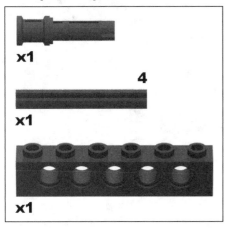

**x1**

**4**

**x1**

**x1**

Put the pin through the brick, and lay the brick on top of the plates. Then insert the axle.

## Bumper Step 2

**x1**

**x3**

Insert the blue pins. The pins on the 1x4 brick will not have enough room to slide all the way through—just push them as far as they will go.

## Bumper Step 3

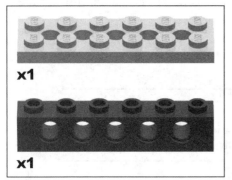

**x1**

**x1**

Attach the brick and plate. You will have to wedge the axle up a little to fit the brick through the pin.

## Bumper Step 4

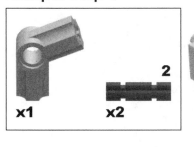

Attach the angle connector and axles. Since this is a pin with friction, the angle of the angle connector is adjustable. For now, align it the way it is shown.

## Bumper Step 5

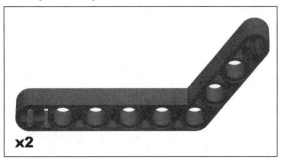

Attach the liftarms. The liftarms will push against each other a little bit, but this is okay.

## Bumper Step 6

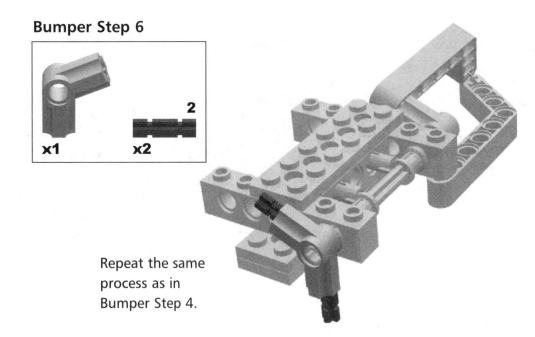

Repeat the same process as in Bumper Step 4.

## Bumper Step 7

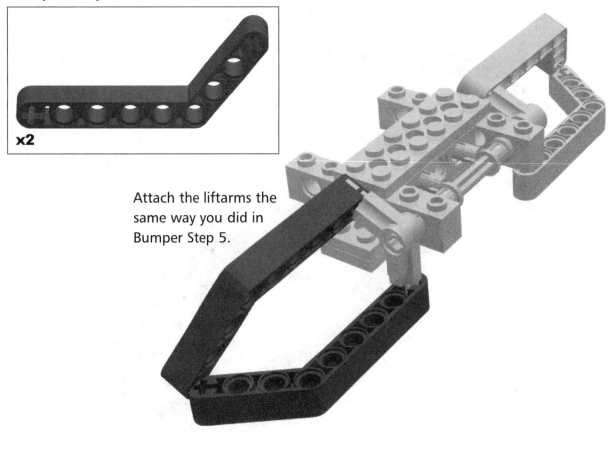

Attach the liftarms the same way you did in Bumper Step 5.

## Bumper Step 8

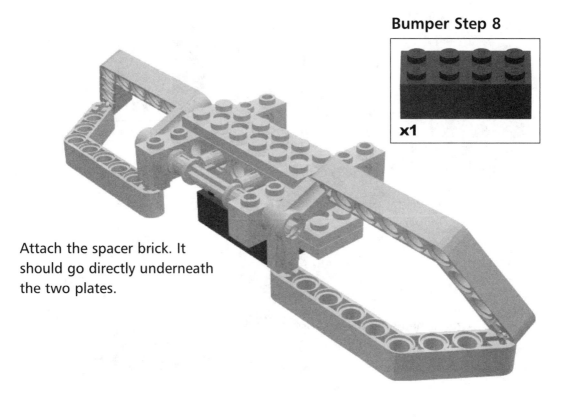

**x1**

Attach the spacer brick. It
should go directly underneath
the two plates.

## Bumper Step 9

**x2**

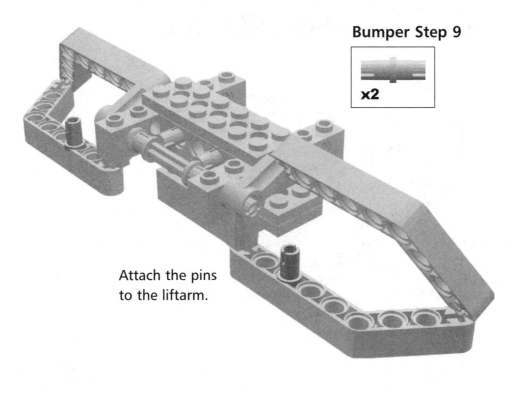

Attach the pins
to the liftarm.

## Bumper Step 10

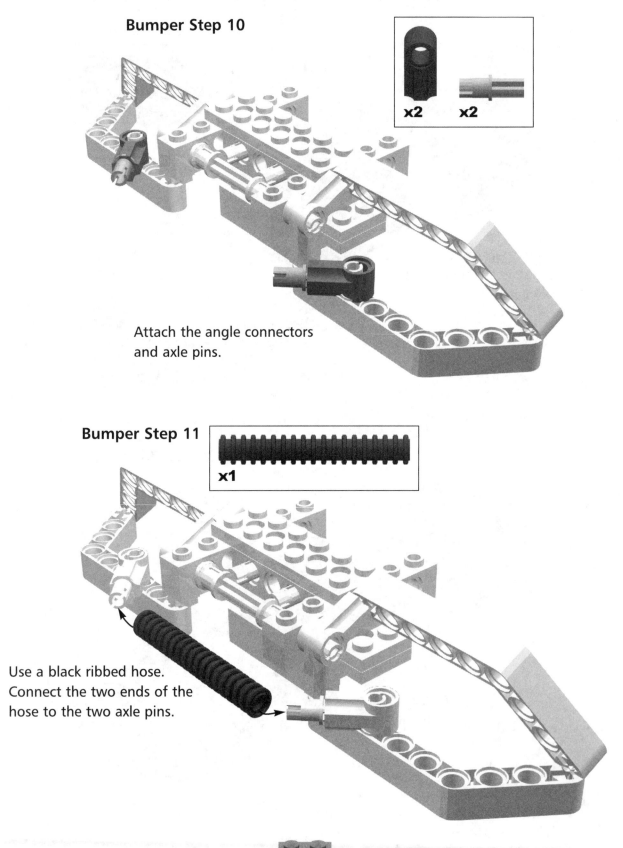

Attach the angle connectors
and axle pins.

## Bumper Step 11

**x1**

Use a black ribbed hose.
Connect the two ends of the
hose to the two axle pins.

## Bumper Step 12

**x4**

Attach four half-length
pins to the liftarms.

### Bricks & Chips...
## Building Tip

This is a common way of setting bricks on top
of horizontal lift arms or other parts with holes.
It connects parts at perpendicular angles so it is
good for bracing in certain situations, too.

## Bumper Step 13

**x2**   **x2**

Seat the 1x2 bricks on top of the
half-length pins. Don't forget to insert
the half-length pins into the 1x2 bricks.

## Bumper Step 14

**x2**

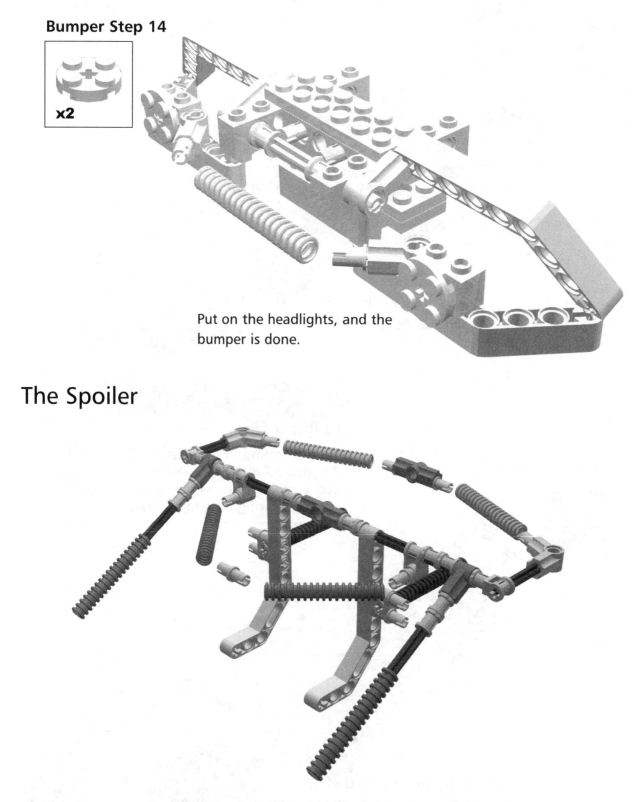

Put on the headlights, and the bumper is done.

# The Spoiler

This is the spoiler and exhaust system. This gives your car the look of an F1 racer.

## Designing & Planning...

## Design Your Own

This is just one spoiler setup. After you build this and see how it works, you can design your own spoiler.

### Spoiler Step 0

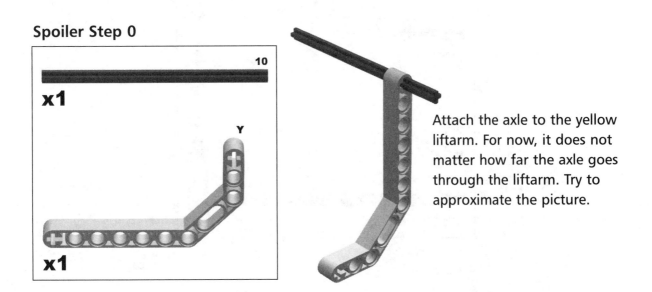

Attach the axle to the yellow liftarm. For now, it does not matter how far the axle goes through the liftarm. Try to approximate the picture.

### Spoiler Step 1

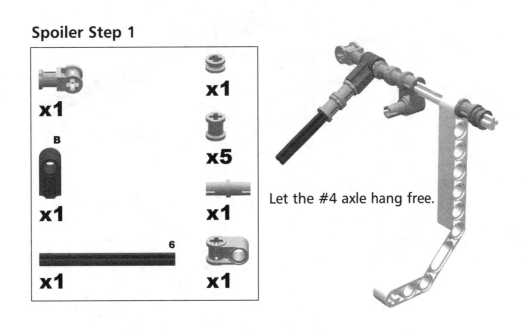

Let the #4 axle hang free.

## Spoiler Step 2

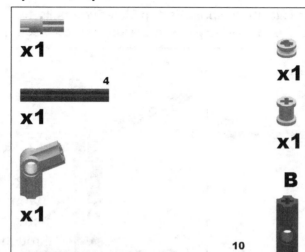

Attach all the parts as shown.

## Spoiler Step 3

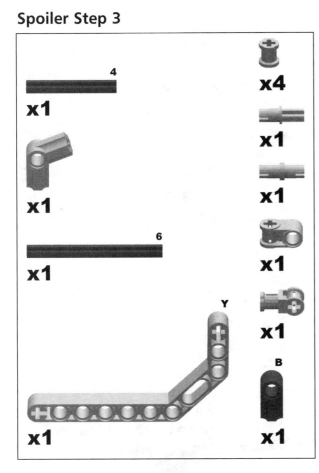

Repeat the same setup you constructed in Spoiler Step 2 and Spoiler Step 3.

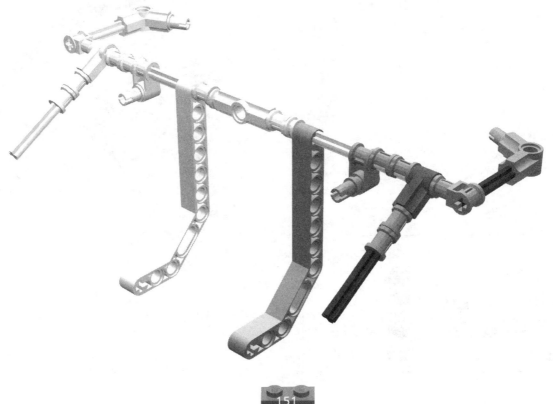

## Spoiler Step 4

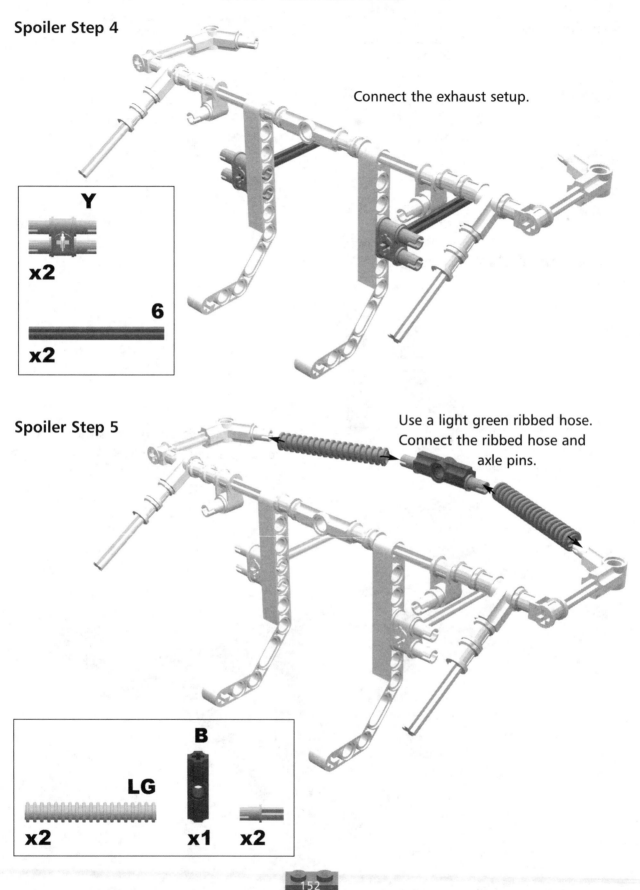

Connect the exhaust setup.

**Y**

x2

6

x2

## Spoiler Step 5

Use a light green ribbed hose.
Connect the ribbed hose and
axle pins.

**LG**

x2

**B**

x1

x2

## Spoiler Step 6

Connect the black ribbed
hose to the axles.

## Spoiler Step 7

Use the magenta
ribbed hose. Connect
the ribbed hose with a
pin in the center.

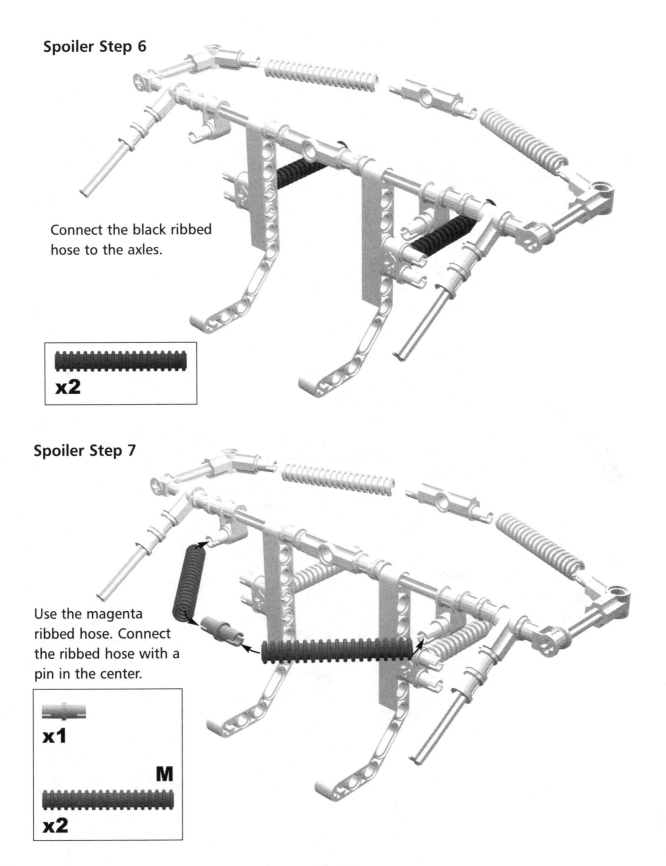

## Spoiler Step 8

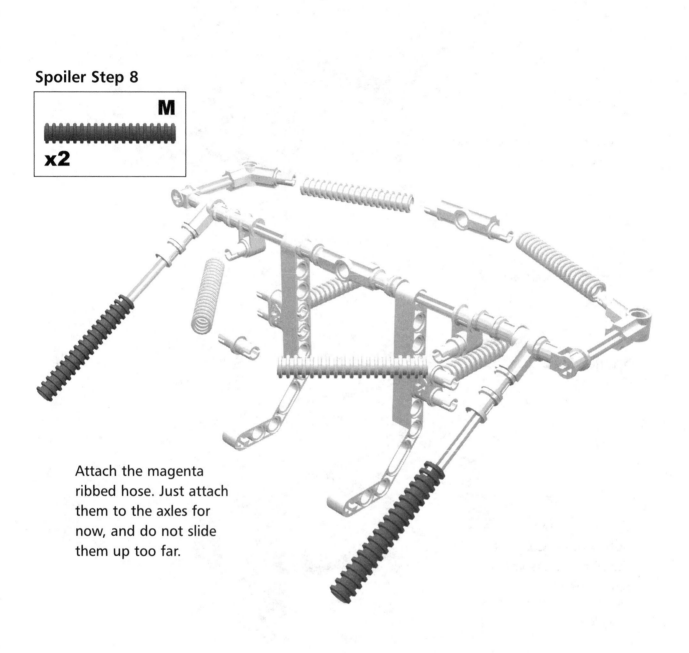

**M**

**x2**

Attach the magenta ribbed hose. Just attach them to the axles for now, and do not slide them up too far.

# Putting It All Together

Finally, put the entire car together. Here you will take all the components that you have built and put them together to complete your F1 Racer!

### Final Assembly Step 0

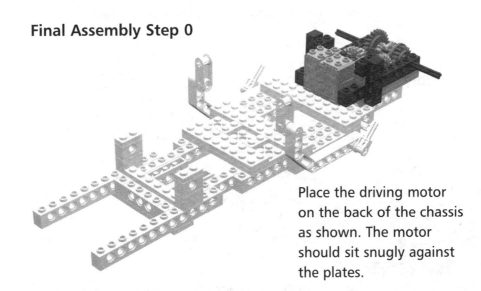

Place the driving motor on the back of the chassis as shown. The motor should sit snugly against the plates.

## Final Assembly Step 1

Place the steering assembly at the front of the chassis. The 1x8 plates should fit directly over the 1x4 brick.

## Final Assembly Step 2

Place the steering motor over the steering assembly. The motor should fit into the spot made for it on the chassis. The plate should go over the 1x2 bricks. The gear should mesh with the rack. This is where the rack and pinion come together. The rotational motion of the motor is translated into horizontal motions of the rack.

## Final Assembly Step 3

Put the bumper on the front of the chassis. The plates of the bumper should fit onto the end of the chassis.

## Final Assembly Step 4

x2

x1

Attach the pins.

## Final Assembly Step 5

x1

Attach the 1x4 liftarm to secure the bumper.

## Final Assembly 6

x2

x1

Turn your model around and attach pins as in Final Assembly Step 4.

## Final Assembly Step 7

x1

Attach the liftarm as you did in Final Assembly Step 5.

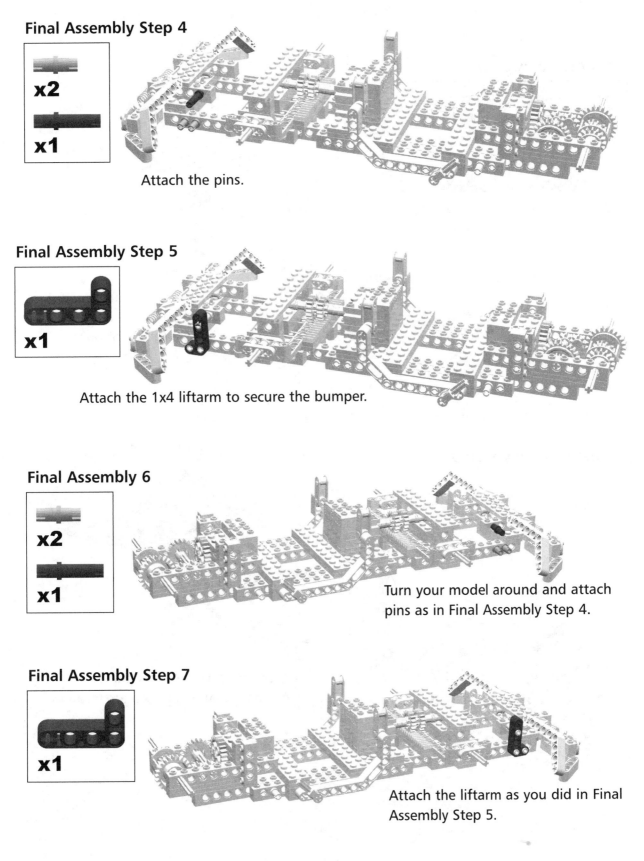

## Final Assembly Step 8

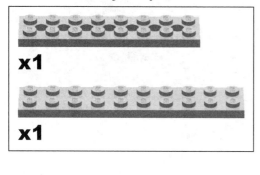

**x1**

**x1**

Flip the model upside down, and brace it horizontally with plates.

## Final Assembly Step 9

Attach the spoiler. At this point, just place the spoiler at the approximate location.

## Final Assembly Step 10

**x1**

Fit the pin into the hole next to the axle hole on the liftarm. The long end of the pin should go through the spoiler to the chassis's last hole. The liftarm should be touching the axle.

## Final Assembly Step 11

8

**x1**

Put the axle through the liftarms and the chassis.

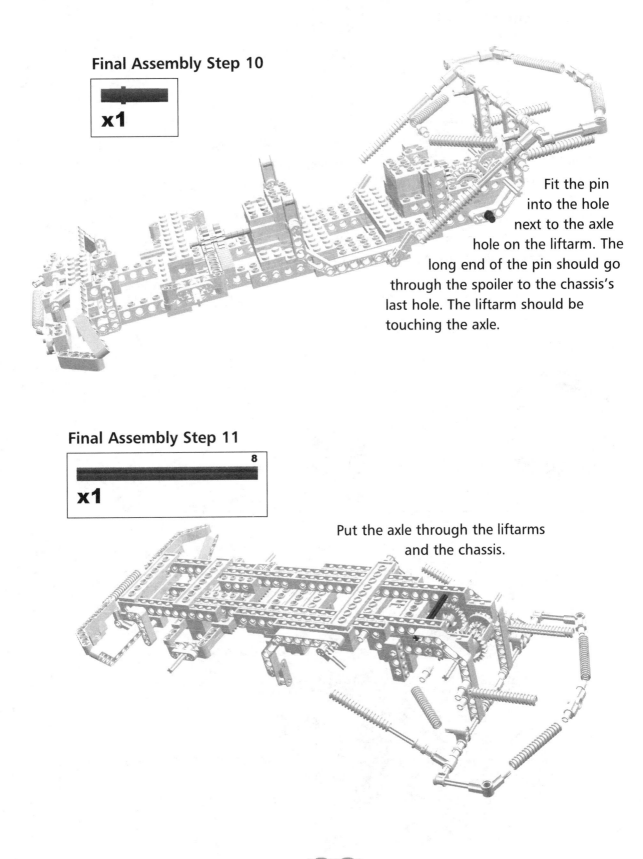

## Final Assembly Step 12

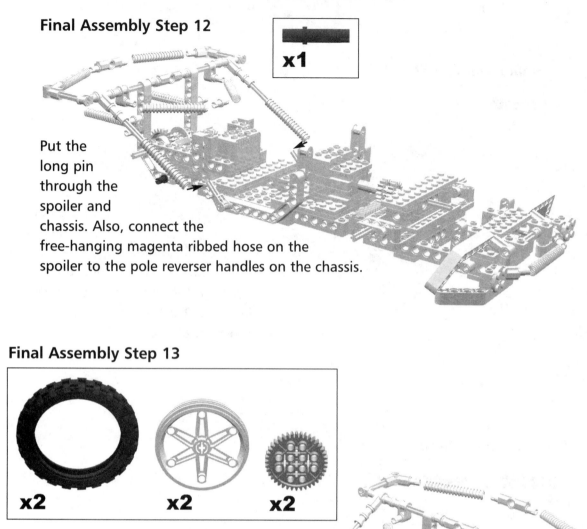

x1

Put the long pin through the spoiler and chassis. Also, connect the free-hanging magenta ribbed hose on the spoiler to the pole reverser handles on the chassis.

## Final Assembly Step 13

x2          x2          x2

Attach the front wheels and hubcaps.

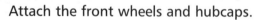

## Bricks & Chips...

## Building Tip

Using double-length pins to go through two parts is a good way to attach parts after they have been constructed. This is useful when building with a modular design.

## Final Assembly Step 14

**x1**

Attach the pin. Make sure the short segment goes into the chassis.

## Final Assembly Step 15

**x1**

Attach the 1x6 brick. One end should go over the axle, while the other should be attached to the pin.

## Final Assembly Step 16

Attach the axle joiner. Then repeat steps Final Assembly Step 13 through Final Assembly Step 16 for the rear wheels.

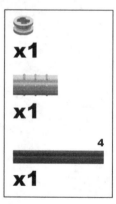

**x1**

**x1**

4

**x1**

## Final Assembly Step 17

**x1**

Set the RCX on the car. Make sure it is securely attached to the plates underneath. The orientation of the RCX does not matter.

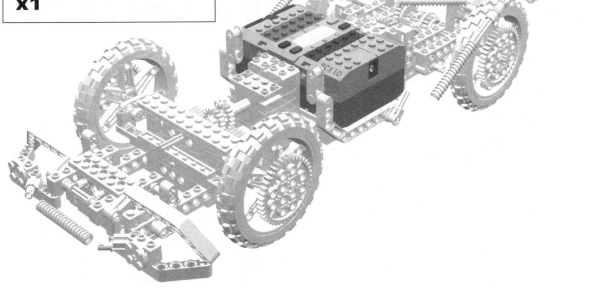

## Final Assembly Step 18

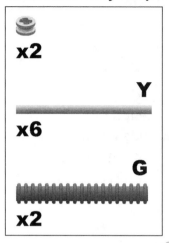

x2

Y

x6

G

x2

Attach the ribbed hose and flex system hose to the free-hanging perpendicular angle connectors, and you're finished! Note that you only need two yellow hoses. Thread the green ribbed hose through the yellow flex system hose and use the half-bushing to secure the green hose. The other end of the hose should go into the blue pins in the bumper.

At this point, you can connect the motors to the RCX and create driving programs! Have fun!

# Robot 5

## The Three-in-One Bot

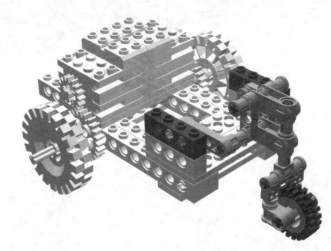

The Three-in-One Bot was built as a simpler yet versatile alternative to the Roverbot that's found in the RIS 2.0 *Constructopedia*. It is a small robot base that can be configured for a variety of functions. It makes a good beginner's model because it uses few parts and can be built within half an hour. In this implementation, it is equipped with a light sensor for line following, for which it is eminently suitable. However, it has the potential to be developed further into a fully functional robot that can be used in competitive challenges such as the FIRST LEGO League (FLL) contests.

The Three-in-One Bot is powered by two of the LEGO geared motors in a differential drive arrangement. The motors are placed in front, creating a sort of front-wheel drive, for better traction and ease of turning. A gear reduction of 3:2 (or 1.5:1) is used to give the robot a balance between speed and reliability when used as a line follower. The robot also features a trailing caster wheel, which gives it great maneuverability, which is an important factor for line following.

The Three-in-One Bot with the light sensor (building instructions for the light sensor variation are available on the Syngress Solutions Web site at www.syngress.com/solutions) will work with most line-following programs. It is customary to have the robot follow the edge of a line (usually the left edge) while it moves forward.

The Three-in-One robot base can be easily customized–we encourage you to try the following changes, observing the effect on the robot's performance:

- **Motor placement** Move the motors back relative to the driving wheels.

- **Gear ratio** Try different combinations of gears.

- **Wheels** Try different types of wheels.

- **Trailing wheel** Replace the trailing caster with other wheel arrangements, such as a sliding pulley wheel.

- **Sensors** Attach bumpers that activate touch sensors, to turn the robot into an obstacle avoidance vehicle.

# The Caster Wheel

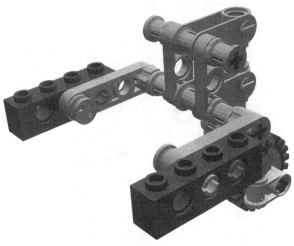

This is the caster wheel sub-assembly that gives the robot its great maneuverability, which is so important for line following.

## Bricks & Chips...

### Building Caster Wheels

There are many ways to build a caster wheel. However, do not use a coupled caster wheel, as Mario Ferrari pointed out in his book *Building Robots with LEGO MINDSTORMS*. For this robot, it's best to use only a single, freely rotating wheel.

## Caster Wheel Step 0

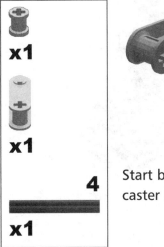

**x1**

**x1**

**4**

**x1**

Start by building the caster sub-assembly.

## Caster Wheel Step 1

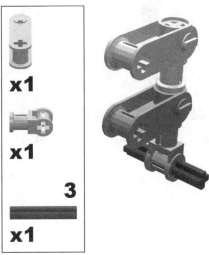

**x1**

**x1**

**3**

**x1**

## Caster Wheel Step 2

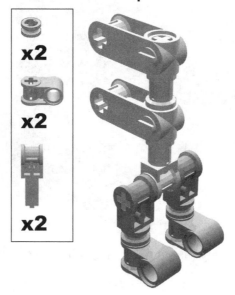

## Caster Wheel Step 3

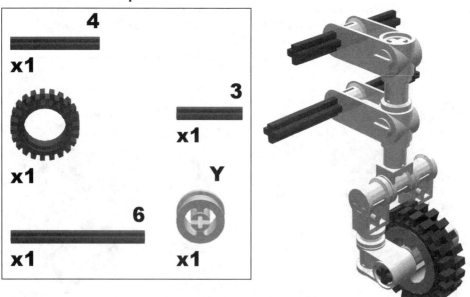

## Caster Wheel Step 4

Make sure the wheel rotates freely on its axle. The caster wheel must also turn a full 360 degrees freely on its vertical pivot.

## Caster Wheel Step 5

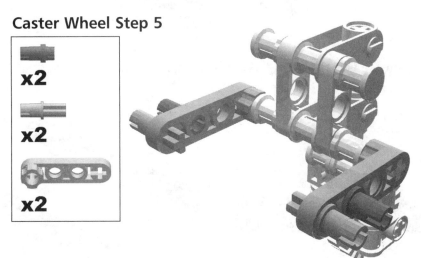

## Caster Wheel Step 6

**G**

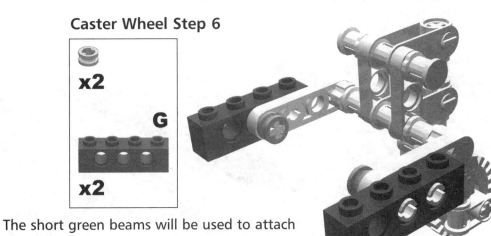

The short green beams will be used to attach the caster sub-assembly to the main robot base.

# Putting It All Together

This is the main Three-in-One Bot robot base, which can be customized for various functions.

## Final Step 0

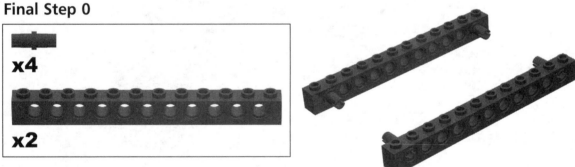

Start by building the side frame of the robot base.

## Final Step 1

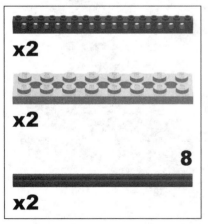

Double up each of the side frames with the long beams. Attach the long plates to the bottom of the beams, then pass the axles through the beams.

## Final Step 2

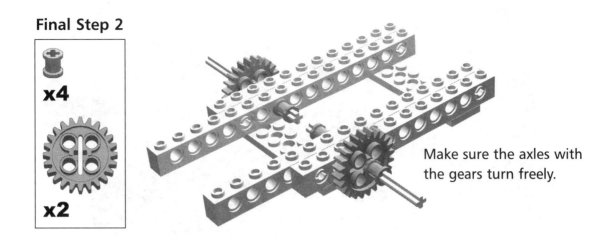

Make sure the axles with the gears turn freely.

## Final Step 3

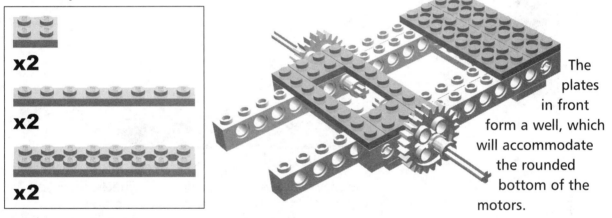

The plates in front form a well, which will accommodate the rounded bottom of the motors.

## Final Step 4

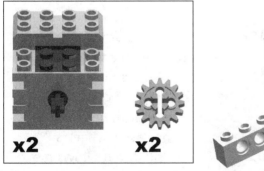

The motor has been raised one plate height to allow the 16t gear to mesh with the 24t gear.

Note that the gear meshing is not perfect, but it is close enough for our purposes.

## Final Step 5

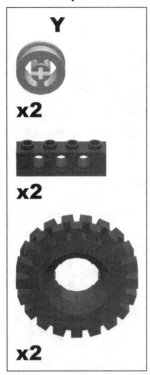

**Y**

x2

x2

x2

Bricks & Chips...
## Using Wheel Variations

You may want to try other types of wheels,
in particular the ones with the big wide tires.

## Final Step 6

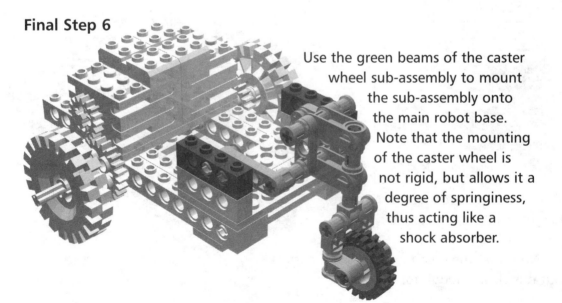

Use the green beams of the caster
wheel sub-assembly to mount
the sub-assembly onto
the main robot base.
Note that the mounting
of the caster wheel is
not rigid, but allows it a
degree of springiness,
thus acting like a
shock absorber.

## Final Step 7

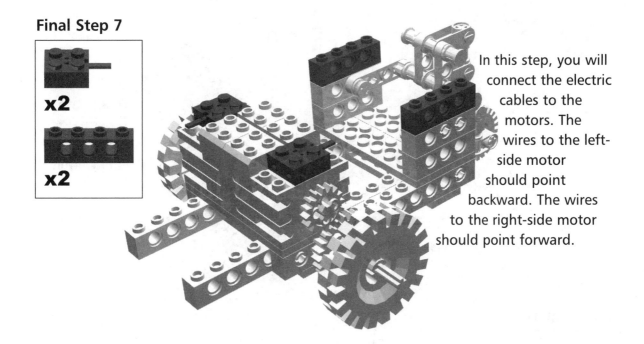

**x2**

**x2**

In this step, you will connect the electric cables to the motors. The wires to the left-side motor should point backward. The wires to the right-side motor should point forward.

## Final Step 8

**x4**

**G**

**x4**

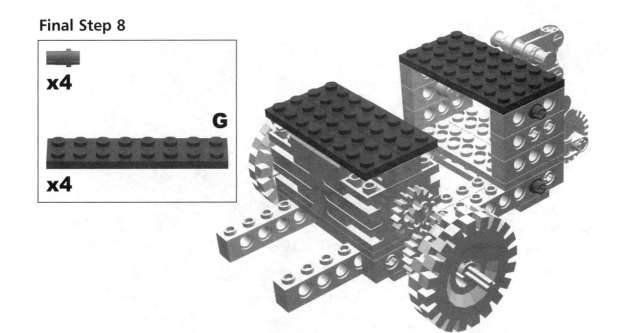

## Final Step 9

Brace the back of the robot base in order to keep the caster sub-assembly firmly in place.

## Final Step 10

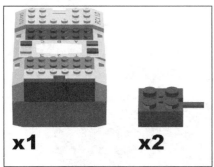

The wires for the right-side motor connect to Port C on the RCX.

Connect the left-side motor wires to Port A on the RCX.

To attach the RCX, first separate the two sections. Put the top section aside. Place the bottom section on top of the green plates, and press down firmly. Replace the top section of the RCX (with the batteries), making sure the IR port faces forward.

Make sure the electric cables are oriented as shown.

## Final Step 11

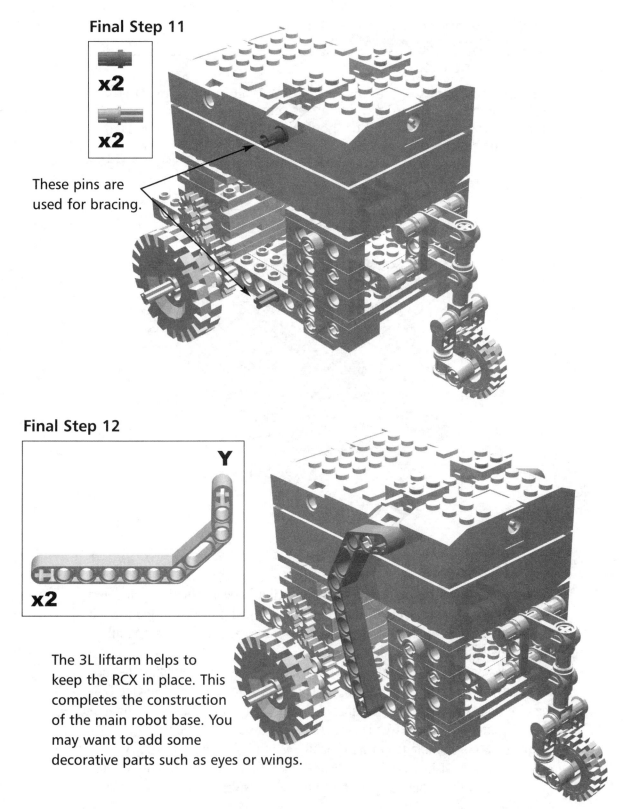

**x2**

**x2**

These pins are
used for bracing.

## Final Step 12

**Y**

**x2**

The 3L liftarm helps to
keep the RCX in place. This
completes the construction
of the main robot base. You
may want to add some
decorative parts such as eyes or wings.

**Bricks & Chips...**

## Testing Your Robot

Test the movements of the robot using RCX built-in program #1. The robot should go forward at quite a fast speed. Next, test it out using built-in program #4. The robot should move in random directions. It should be able to turn freely on its caster wheel.

## SYNGRESS
syngress.com

# Adding a Light Sensor

With the addition of a light sensor attachment, the Three-in-One Bot will detect and follow lines on a playing field such as the test pad provided in the RIS 2.0. Line following is an important task in many competitive challenges, and all good robots must be able to do it reliably.

The designer of the Three-in-One Bot has created a set of building instructions that will show you how to add the light sensor to your Three-in-One Bot. The directions for building an additional light sensor sub-assembly can be found on the Syngress Solutions Web site at www.syngress.com/solutions.

To check out the line following capabilities of the Three-in-One Bot we encourage you to visit www.geocities.com/laosoh/robots/linefollow.htm, where the designer of this robot has posted some of the results and pictures of the Three-in-One Bot in action.

# Robot 6

## The Aerial Tram

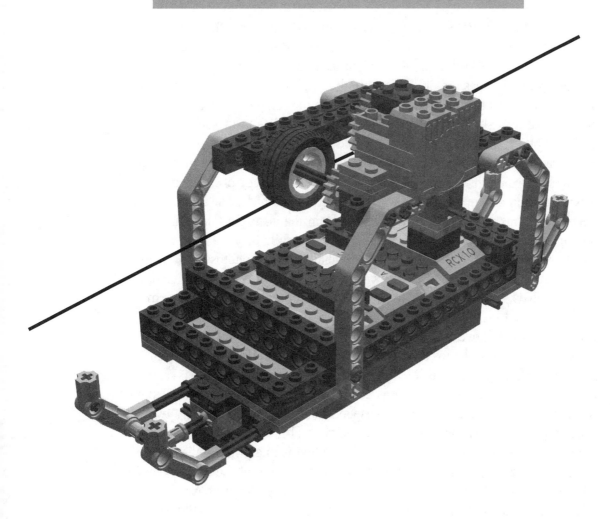

The LEGO Aerial Tram is modeled after the Aerial Tram that glides over the Arkansas River, slicing through the granite canyon of the Royal Gorge in Colorado. Trams are built to traverse expanses that are often at great heights or in areas where bridges are difficult to build. Aerial trams can have many purposes–some trams are for sightseeing, some are for cargo or passenger transportation such as ski lifts, and some are built to complete tasks such as dropping off or picking up cargo that would be impossible to transport any other way. The Rain Forest Tram in Costa Rica, for example, was built to take people sightseeing through the different levels of the canopy of the rainforest–a trip that would be impossible with any other type of vehicle. The Palm Springs Aerial Tram was constructed to take people from the heat of a California day to the top of the snow-capped cliffs of Chino Canyon while they experience a 30-degree temperature change (read technical details and see photos of the Palm Springs Aerial Tram at  www.pstramway.com/history-tech/technical-detail.html).

The LEGO Aerial Tram can be customized and programmed to perform many tasks and respond to sensor input. The basic tram design is powered by one motor, which moves the tire over the cabling wire from which the tram is suspended. To create the tower supports and cabling necessary for the tram to fly, tie a medium-gauge wire (22-gauge works well) to the ends of two chair backs, placing books on each chair as counterweights for the tram (or substitute two other suitable tower supports to suspend the cable wire). There are two ways to power a tram. One method is to power and move the cable, allowing the trams to move with the cable. Another method is to attach a pulley or wheel to the tram, powering the pulley with a motor so that the tram traverses a stationary cable. The LEGO Aerial Tram is powered in this way. A gear is attached to a motor shaft that moves a tire along a stationary cabling wire. There are two touch sensor bumpers at each end that can be programmed to beep, pause, and reverse the tram's direction. A light sensor is mounted on the main cab and can be programmed to control the tram's movements over the cabling wire by responding to electrical tape folded over in small flags on the wire. Another way of using the light sensor for control is to program it to respond to the beam from a flashlight when the beam is pointed directly at the light sensor. This beam could be a bad-weather warning signal that tells the tram to come back to the tower.

On each end of the tram, a bench is ready to seat your favorite LEGO figures for transport. The bottom of the tram allows for the cover of the RCX to be removed easily for battery replacement. To enhance your LEGO Aerial Tram further, you could make attachments to add to the bottom of the tram. A gondola attachment could carry cargo in addition to passengers. A grabber arm could be built and powered with the second motor for picking up or dropping off items at certain locations or for completing tasks too hard to reach. Begin with the basic design shown here, and let your imagination run wild as your Aerial Tram flies over the canyon of your choice!

# The Aerial Tram Base

The tram base secures the RCX and provides the foundation for the rest of the tram assemblies. A light sensor is located on the tram base and can be programmed to control the tram's movements.

## Tram Base Step 0

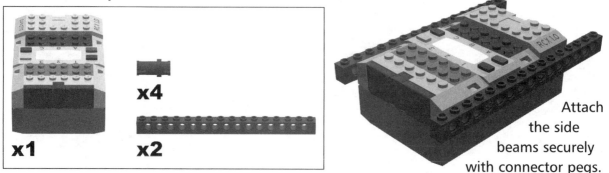

Attach the side beams securely with connector pegs.

## Bricks & Chips...

### New Batteries

The Aerial Tram's base is built for easy removal of the RCX cover when changing batteries. Planning for battery replacement is an important part of building a robot.

## Tram Base Step 1

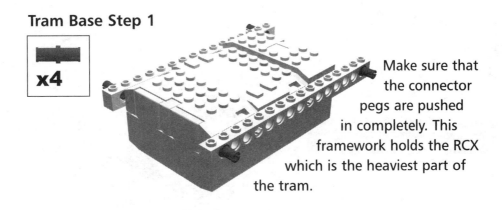

**x4**

Make sure that the connector pegs are pushed in completely. This framework holds the RCX which is the heaviest part of the tram.

## Tram Base Step 2

**x2**

**x2**

## Tram Base Step 3

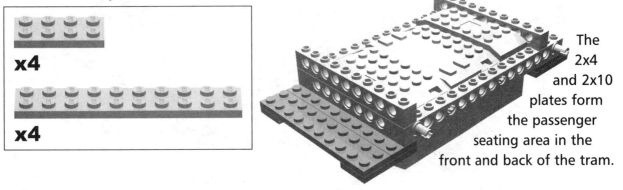

**x4**

**x4**

The 2x4 and 2x10 plates form the passenger seating area in the front and back of the tram.

## Inventing...

## Passenger Seating

The tram is now ready to carry your favorite LEGO figures to the other side of the canyon. If they are workers, they might be carrying tools. If they are tourists, they might hold cameras.

## Tram Base Step 4

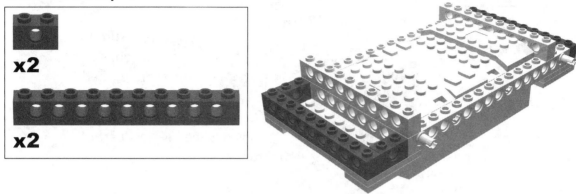

x2

x2

### Inventing...

## Gondola

Build a gondola and attach it to the bottom of the tram to carry cargo across the canyon.

## Tram Base Step 5

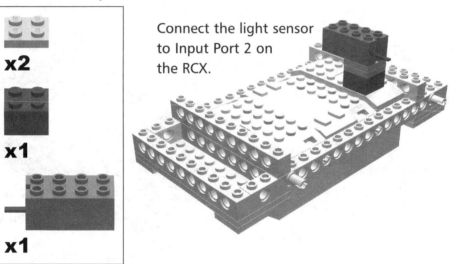

x2

x1

x1

Connect the light sensor to Input Port 2 on the RCX.

### Developing & Deploying...

## Input Ports

Make sure that the tram program is written for a light sensor attached to Input Port 2. This will be more important later as you add touch sensors to the other RCX input ports.

## Tram Base Step 6

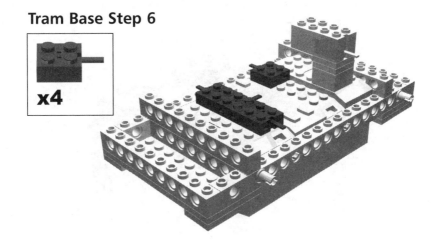

**x4**

Developing & Deploying...

## Light Sensor Ideas

Wrap small flags of black electrical tape around the cabling wire that the tram will traverse. Program the tram to detect the tape, beep a signal, and pause for a sightseeing stop so that passengers can take pictures of the view before finishing the route.

Inventing...

## Flashlight Control

Program the tram to stop and play a warning series of beeps when the light sensor detects the bright light from a flashlight. The flashlight could be a warning beacon to signal high winds or bad weather, thus sending the tram back without making any more photo stops.

# The Cab

The cab provides the pulley and wheel system that is used by the tram to traverse the cable wire.

## Cab Step 0

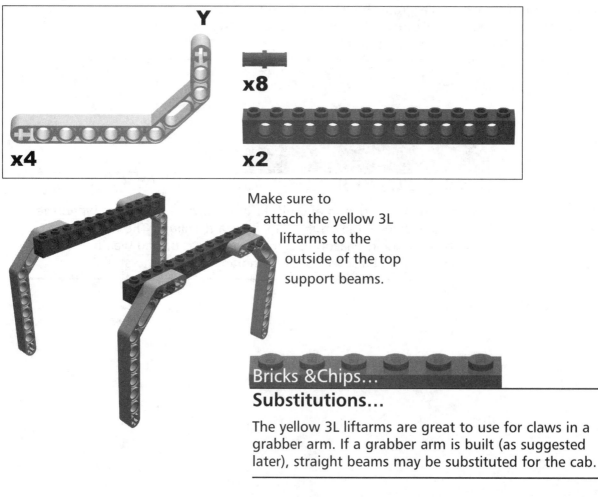

Y

x8

x4

x2

Make sure to attach the yellow 3L liftarms to the outside of the top support beams.

Bricks &Chips...

## Substitutions...

The yellow 3L liftarms are great to use for claws in a grabber arm. If a grabber arm is built (as suggested later), straight beams may be substituted for the cab.

## Cab Step 1

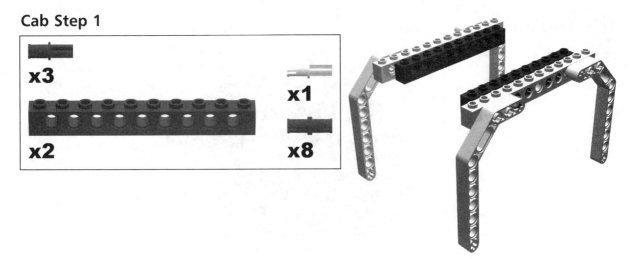

**x3**

**x1**

**x2**

**x8**

## Cab Step 2

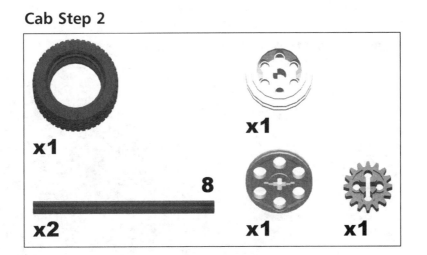

**x1**

**x1**

**8**

**x2**

**x1**

**x1**

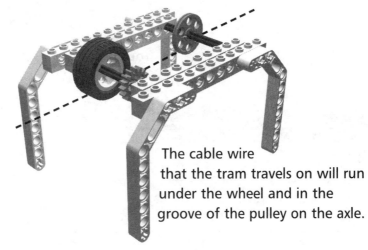

The cable wire that the tram travels on will run under the wheel and in the groove of the pulley on the axle.

# The Motor

The tram uses only one motor. Plug the electrical wire into RCX Output Port B.

## Motor Step 0

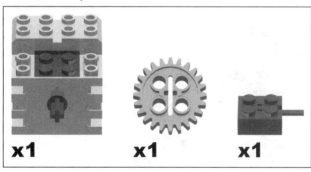

**x1**          **x1**          **x1**

The gear provides the power from the motor to the wheel.

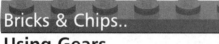

Bricks & Chips..

## Using Gears

Gearing up (driving a smaller gear with a larger gear) provides speed. Gearing down (driving a larger gear with a smaller gear) provides more power or torque.

## Motor Step 1

Make sure that the motor is attached to RCX Output Port B and that the program is written to power Port B.

### Inventing…

## Grabber Arm

Use the second motor to power a grabbing arm. Attach the arm to the base of the tram. The arm can carry things across the canyon, pick things up, or drop items where needed. Program the light sensor to open and close the grabber arm in response to a flashlight signal.

# The Touch Bumper

The bumpers on each end of the tram allow the tram to sense when it is at the end of the cable line, so it can stop or pause and reverse directions. You will need to build one bumper for the front of the tram and another one for the back.

## Bumper Step 0

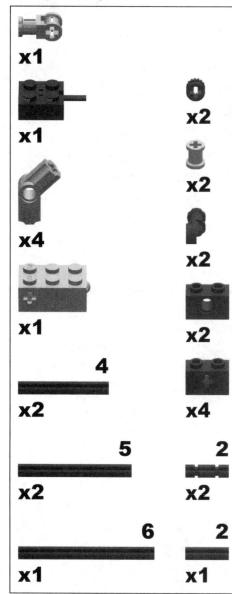

Each bumper will hold a touch sensor. Attach the electrical wires to RCX Input Ports 1 and 3. Make sure that the program is written for the touch sensors connecting to these ports. Remember to build two bumpers!

### Bricks & Chips...
### Touch Bumpers

It is important to create a bumper in front of the touch sensor instead of using the touch sensor by itself. The bumper increases the surface area of the touch sensor and allows for a better response. It is also good to create the bumper so that different amounts of pressure will allow contact to the touch sensors. Rubber bands or flexible tubing pieces are also good choices when building bumpers. The Constructopedia has several different bumper ideas that could be substituted here.

### End of the Cable Line

Program the RCX to beep, pause for passenger unloading, and then reverse direction when the touch sensors are depressed.

## Putting it All Together

All of the sub-assemblies are built and are ready to be put together. This last step will put the final touches on the Aerial Tram.

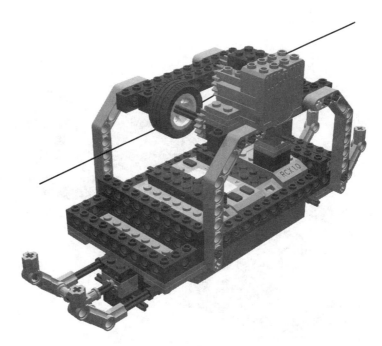

Make sure that the electrical wires are plugged in to the assigned RCX ports and all pieces are snapped together firmly.

■ Attach the light sensor wire to Input Port 2.

■ Attach the motor output wire to Output Port B.

■ Attach the front bumper touch sensor wire to Input Port 1.

■ Attach the back bumper touch sensor wire to Input Port 3.

Make sure that each end of the cable wire is securely tied to the back of chairs or "towers," and that the "towers" secure enough to offset the weight of the RCX traveling on the cable wire. Place the Aerial Tram on the wire in the groove of the pulley, and align the wheel.

## Developing & Deploying...

# Remote Control

Use only one touch bumper. Attach the other touch sensor to a long electrical wire to create a remote control. Connect one end of the electrical wire to the RCX input port, and connect the other end to the touch sensor. This will create a wired remote control. Program the touch sensor to stop when depressed and start again in reverse when released. Add a beep or two to the program. This remote control could also control a grabber arm.

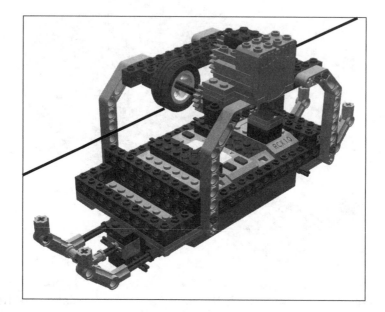

# Robot 7

## The LEGO Safe

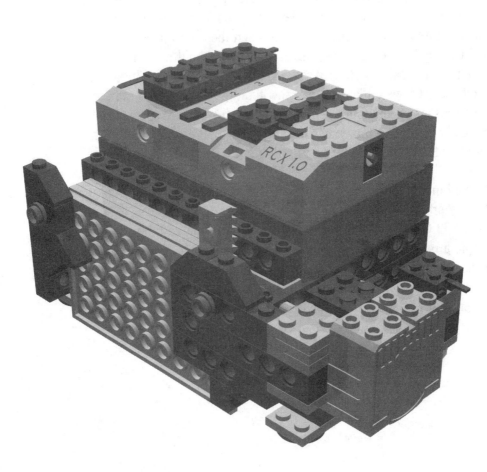

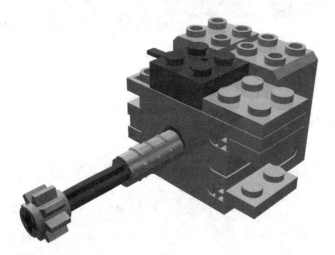

The vault area of the LEGO Safe is the perfect place to keep valuables protected from intruders. The Safe is automated and requires a special sequence to be pushed into the touch sensors in order for the door to open. The light sensor hidden inside the Safe will sound an alarm when the door opens. Additional LEGO bricks may be used to make the Safe bigger to hold more items.

Unlike the bank safes seen in Westerns, the LEGO Safe is computer controlled and doesn't have a dial. To open the door, the owner must remember a sequence of pushes on the touch sensors. The touch sensors are color coded with colored bricks to aid the owner in remembering the code. If someone snoops and finds out the code, re-programming the RCX can easily secure the Safe again.

You could invent variations on the Safe; for example, you could make it a music box by programming the light sensor to play a tune instead of sounding a warning. A LEGO figure attached to the end of a bendable tube could pop out and spin around using the power from a second motor–just like the ballerina inside a beautiful music box!

Follow the instructions in this chapter, adding your own final touches, to create a personal LEGO Safe to keep belongings secure.

# The Vault

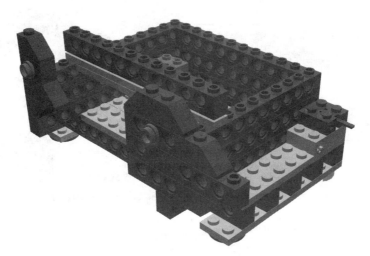

The vault is the storage area for the Safe. The two touch sensors are located in the back of the vault. They must be pushed in a particular sequence determined by your program in order for the Safe's door to open. The light sensor is also built inside the vault and can be programmed to sound an alarm when the door is opened.

## Vault Step 0

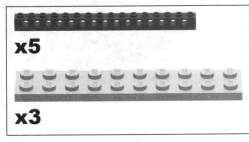

**x5**

**x3**

Begin building the base of the vault. Snap the bricks together securely.

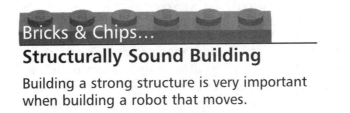

Bricks & Chips...

## Structurally Sound Building

Building a strong structure is very important when building a robot that moves.

## Vault Step 1

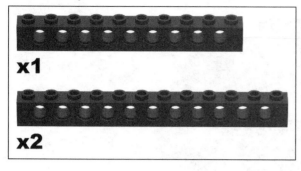

**x1**

**x2**

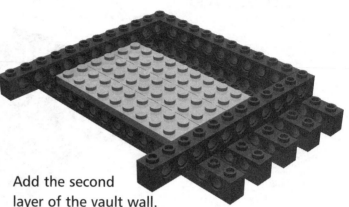

Add the second
layer of the vault wall.

## Vault Step 2

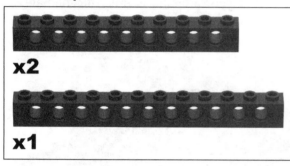

**x2**

**x1**

Continue to build up
the walls of the vault.

### Inventing...
## Additional Vault Storage

To increase storage room in the Safe, add more LEGO
bricks and continue building layers to the vault wall.
Increase the door size by attaching another plate as
an extension. Extend the track that the door travels
on by adding the appropriate number of track pieces.

## Vault Step 3

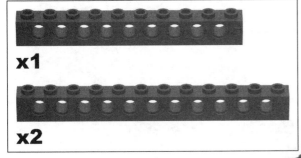

**x1**

**x2**

Add the final beams to
finish the vault walls.

## Vault Step 4

**x1**

**x3**

Add these bricks to begin building
the track for the door.

Inventing...

## Building a Music Box

To create a LEGO music box, program the light sensor
to play a tune instead of sounding an alarm.

## Vault Step 5

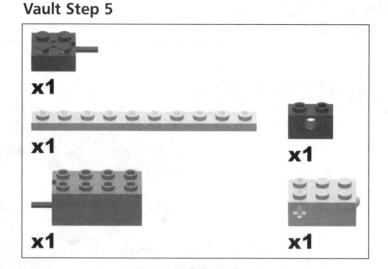

Add one touch sensor to the back of the Safe. Place the light sensor inside the vault. String the electrical wire from the light sensor through the back gap in the floor of the vault.

Inventing...

## Creating a Key System

Create a LEGO key system that would allow the touch sensors to respond to the bumps and indentations on a LEGO key. This would work just like an actual car key!

## Vault Step 6

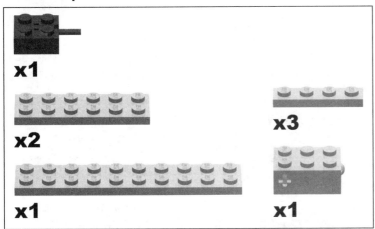

**x1**

**x2**

**x1**

**x3**

**x1**

Add the second touch sensor to the back of the Safe. Add more framing for the door.

## Color Coding the Touch Sensors

Use 2x2 rounded plates under each touch sensor that are different colors. The colors will identify each touch sensor so that the touch sequence necessary to open the vault is easier to remember.

## Vault Step 7

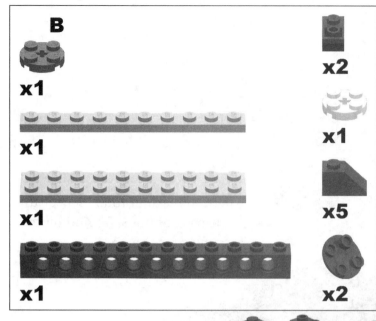

Finish the framing for the Safe door. Add skid plates on the bottom front of the Safe. Add the 2x2 rounded plates under the touch sensors at the bottom rear of the Safe.

## Vault Step 8

Inventing...

## Caught Red Handed

Program the door to close automatically after it has been opened for a set period of time. If an intruder is entering the Safe, he will be caught unaware.

# The Motor

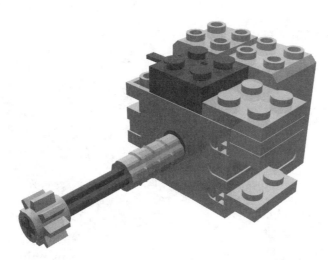

The motor engages and opens the Safe door when the correct sequence is pushed into the touch sensor pads.

### Motor Step 0

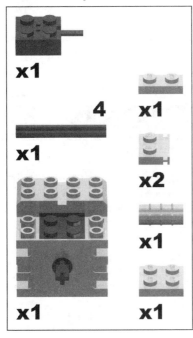

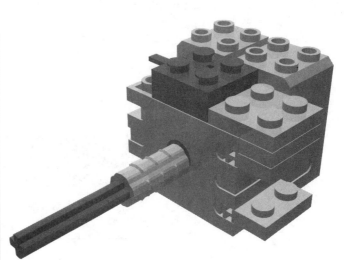

Put the motor sub-assembly together and attach it to the vault before placing the gear on the end of the axle.

**Bricks & Chips...**

## Using an Axle Extender

The axle extender is a good tool to use when an axle needs to be lengthened.

**Motor Step 1**

**x1**

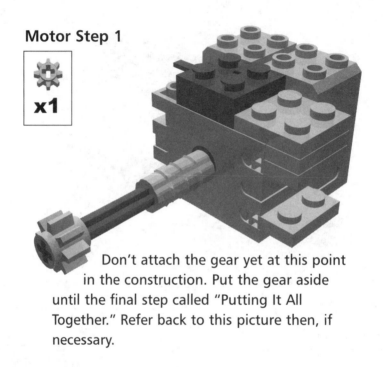

Don't attach the gear yet at this point in the construction. Put the gear aside until the final step called "Putting It All Together." Refer back to this picture then, if necessary.

**Developing & Deploying...**

## Changing the Safe Code

The touch sensor sequence necessary to open the Safe can be re-programmed at any time.

# The RCX

The RCX is the brain of your LEGO Safe.

## RCX Step 0

Attach the motor to Output Port A.

**x1**        **x1**

Developing & Deploying...

## Programming Sensors

Make sure your program is written for the correct sensors and motor and their corresponding ports.

## RCX Step 1

**x3**

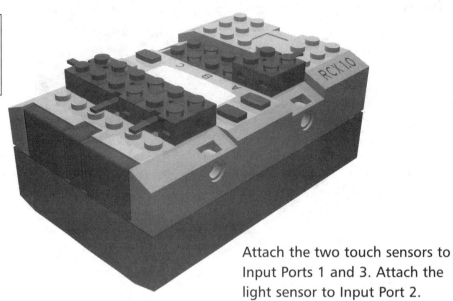

Attach the two touch sensors to Input Ports 1 and 3. Attach the light sensor to Input Port 2.

# The Door

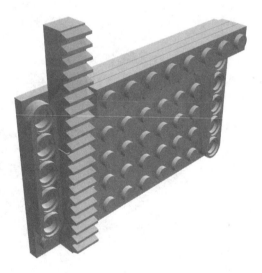

The Safe door moves up and down on the track after the correct code has been pushed into the touch sensors.

## Door Step 0

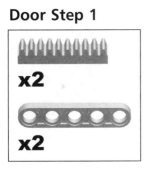

x2

x5

x1

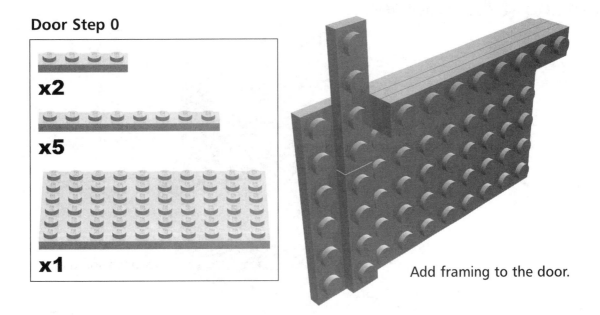

Add framing to the door.

## Door Step 1

x2

x2

Add the track
system. Push the 1x5
plate with holes firmly into
the back of the door. The door will
move up and down on the track when
the motor turns the gear.

# Putting It All Together

The LEGO Safe is ready
to stash some cash!

Firmly attach all of the sub-assemblies together. Plug the electrical wires into the corresponding ports. Add the gear on the end of the motor axle (refer to Motor Step 1). Align the gear with the track on the door.

# Robot 8

## The ULK
## (Useful LEGO Knowledge)

The ULK is a differential drive robot with an arm that can pick up small objects like LEGO bricks. Using only two motors, ULK can drive forward, turn, raise and lower its arms, and open and close its fingers. All of this is made possible by some sophisticated mechanical design. A ratchet splitter uses one motor for both locomotion and steering. A clever lifting mechanism allows a second motor to perform all the arm related functions.

ULK is based on a robot I saw in Jonathan Knudsen's book, *The Unofficial Guide to LEGO MINDSTORMS Robots*. Jonathan's robot, named Minerva (www.oreilly.com/catalog/lmstorms/building/minerva1.5) after the ancient roman goddess of wisdom, performed all of the same actions that ULK is capable of, however ULK performs these actions using different mechanisms. Jonathan's Minerva was, in turn, based on Ben Williamson's FetchBot (http://ozbricks.com/benw/lego/fetchbot/index.html). Jonathan modified Ben's three motor design so it could be built using only the parts in the Robotics Invention System (RIS) 1.0 kit. Unfortunately, due to changes in the RIS parts inventory, Minerva cannot be built by solely using neither the RIS 1.5 nor the RIS 2.0 sets. ULK is the latest in the FetchBot and Minerva ancestry, and can be built using only the parts available in RIS 2.0 set.

# The Arm

ULK wanders around a room searching for LEGO bricks, which it uses to build its nest. When the light sensor spots a brick, ULK picks up the brick, and then returns to its nest to deposit it.

## Arm Step 0

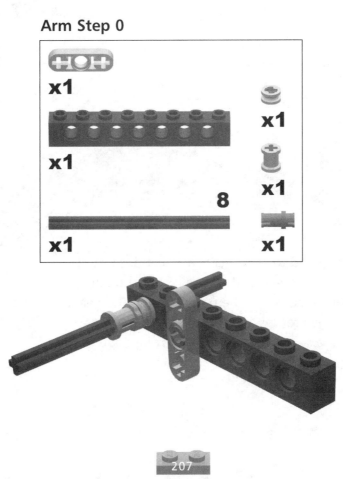

## Arm Step 1

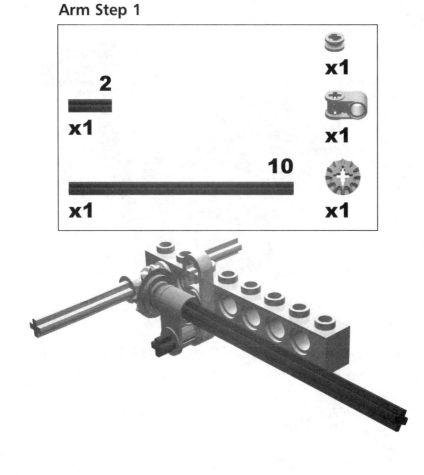

## Arm Step 2

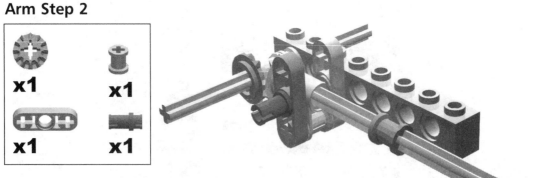

All of the force required to raise ULK's arm is transmitted through the two bevel gears at the base of the shoulder. This generates lots of stress on the parts holding the gears in place. The strange bracket built out of the 3L liftarms and perpendicular axle joiner is definitely strong enough to hold the bevel gears together.

## Bricks & Chips...

# LEGO Connections

LEGO connections are really strong in compression (the force that pushes the bricks together) and in shear (the force that tries to slide the bricks across each other), but they are relatively weak in tension (the force that tries to pull them apart).

When I first built ULK, I supported the #10 axle on both ends using 1x2 beams reinforced with plates on the top and bottom. The snap-on connections proved too weak (in tension) to hold the bevel gears together when ULK tried to raise its arm. The improved support bracket is attached directly to the side beams. The large forces that used to tear the arm apart are now in shear across the three-quarter length pins. The pins can easily counteract the forces, allowing ULK to raise and lower its arm without concern.

## Arm Step 3

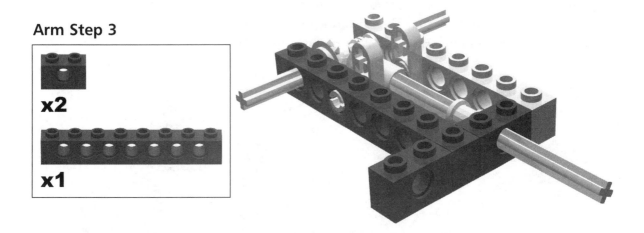

**x2**

**x1**

## Arm Step 4

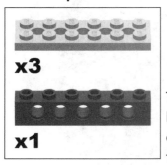

**x3**

**x1**

The double-high stack of plates on the bottom of the arm provides mounting holes for ULK's light sensor.

## Arm Step 5

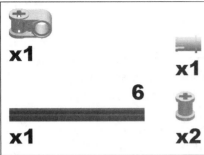

x1

x1

6

x1

x1

x2

The #6 axle acts as a stop to prevent ULK's arm from dropping too far when reaching down to pick up LEGO pieces. The perpendicular axle connector and half-length pin act as part of a stop that prevents ULK from raising its arm too high.

Inserting the #6 axle is tricky. Interference between the ridges on the bushings and the studs on top of the beams prevent you from attaching the bushings after the axle is in place. Instead, first lay the bushings in approximately the right position, and hold them in place as you slide the axle through.

## Arm Step 6

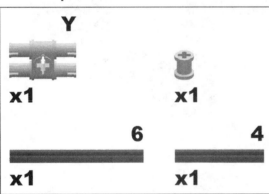

Y

x1

x1

6

4

x1

x1

The yellow double pin and the #4 axle are part of a mounting bracket for the light sensor. The double pin doesn't snap all the way into the bottom of the 2x6 plates, but the connection is strong enough to support the sensor.

## Arm Step 7

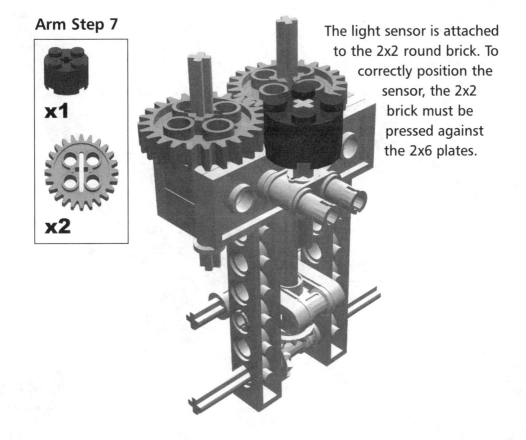

**x1**

**x2**

The light sensor is attached to the 2x2 round brick. To correctly position the sensor, the 2x2 brick must be pressed against the 2x6 plates.

## Bricks & Chips...

## Aligning the Fingers

It is important for the 24t gears to be aligned properly, otherwise the fingers won't be able to pick up LEGO bricks. Adjust the gear position such that the elongated slots in the gear face are almost parallel.

## Arm Step 8

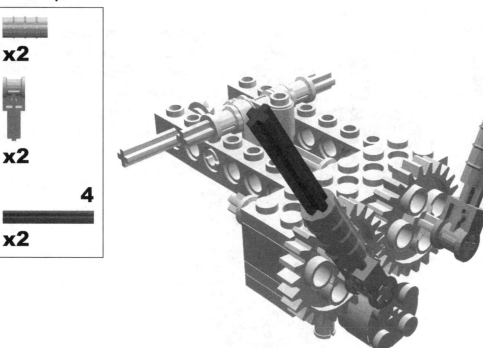

x2

x2

4

x2

## Arm Step 9

Y

x2

x2

x1

The #4 axles make ULK's fingers just a little too long. They scrape the tires on the ground when ULK reaches down to pick something up. Because #3 axles are a little too short and cause ULK to drop things, ULK uses the longer axles, but positions the grabbers about a half-stud away from the end of the axle.

# The Lifter

The lifter returns ULK's arm to the upright position after it drops off a brick.

### Lifter Step 0

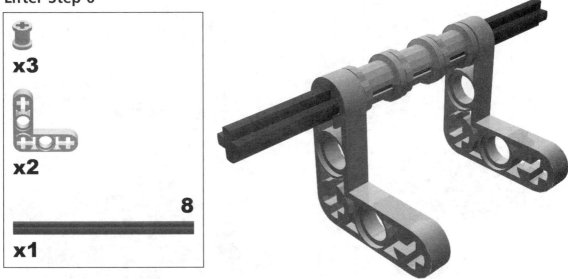

# The Limit Switch

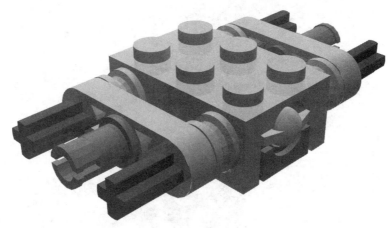

The Limit Switch notifies ULK when its arm is in the upright position. Attempting to drive ULK's arm past the upright position will damage the robot.

### Limit Switch Step 0

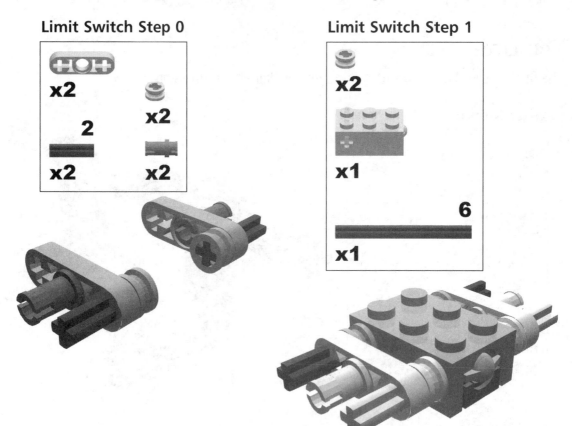

x2

x2

2

x2

x2

### Limit Switch Step 1

x2

x1

6

x1

## Arm Motor

The Arm Motor sub-assembly houses the motor used to raise and lower ULK's arm. It also contains the Limit Switch sub-assembly that signals when ULK's arm is in the upright position.

## Arm Motor Step 0

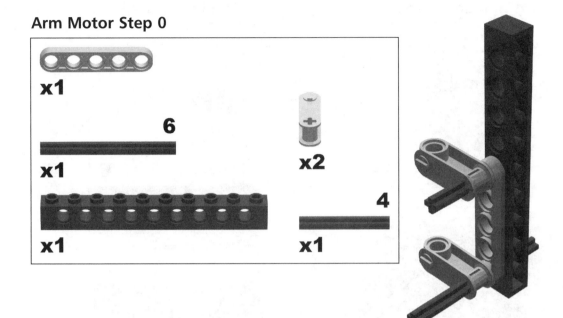

## Arm Motor Step 1

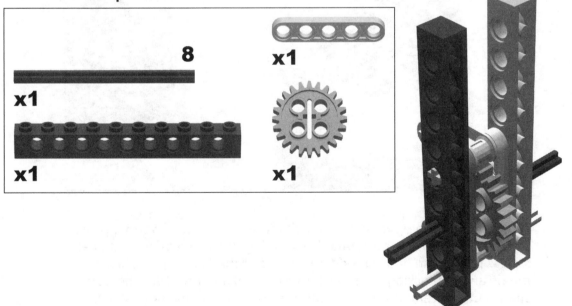

## Arm Motor Step 2

## Arm Motor Step 3

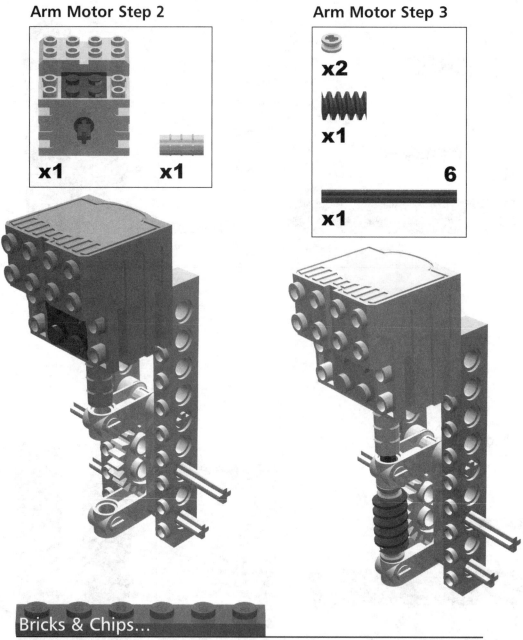

x1          x1

x2

x1

6

x1

## Bricks & Chips...

## Worm Gears

The efficiency of a worm gear is much lower than that of normal gears because the worm works primarily by sliding, thus increasing frictional losses. This has an unusual side effect in that the worm gear is asymmetric and self-locking. You can turn the input shaft to drive the output shaft, but you cannot turn the output shaft to drive the input shaft. ULK uses the worm gear's self-locking feature to hold the arm in the raised position.

## Arm Motor Step 4

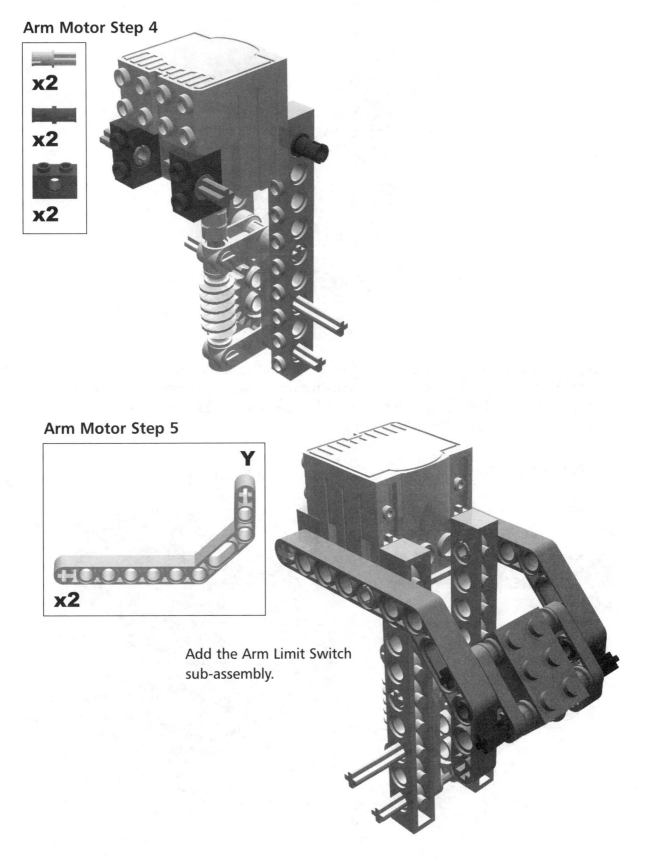

## Arm Motor Step 5

**Y**

x2

Add the Arm Limit Switch
sub-assembly.

# Arm Assembly

ULK's arm is controlled by a single motor. When the motor turns clockwise (reverse direction according to the RCX), ULK lowers its arm, closes the fingers, and then raises the arm. Running the motor in the counter-clockwise direction (forward direction) lowers ULK's arm, opens the fingers, and then returns the arm to the upright position.

## Arm Assembly Step 0

**x8**

**x2**

## Arm Assembly Step 1

**x1**

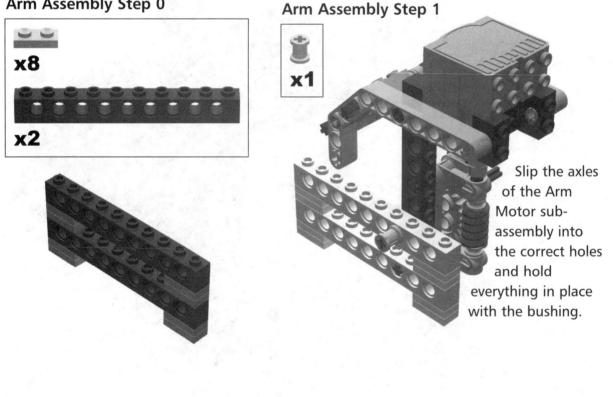

Slip the axles of the Arm Motor sub-assembly into the correct holes and hold everything in place with the bushing.

## Arm Assembly Step 2

**x1**

Add the Lifter sub-assembly. The lifter raises ULK's arm when the motor rotates in the counter-clockwise direction. The fingers (3x3 liftarms) push against the bottom of the neck, returning the arm to the upright position. Spinning the motor clockwise causes the lifter to rotate out of the way.

## Arm Assembly Step 3

**x1**

In this step we will add the Arm sub-assembly built previously. Turning the motor in the clockwise direction causes the fingers to close. Gripping something in the fingers prevents gears in the arm from turning further. This generates a large amount of torque on the arm shaft, causing the arm to rise.

## Arm Assembly Step 4

**x8**

**x2**

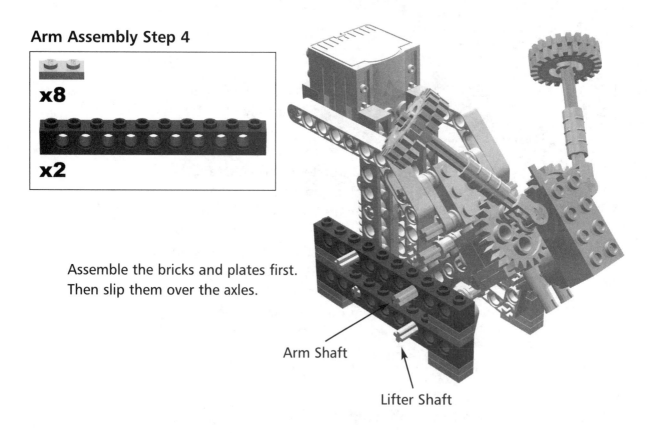

Assemble the bricks and plates first.
Then slip them over the axles.

Arm Shaft

Lifter Shaft

## Arm Assembly Step 5

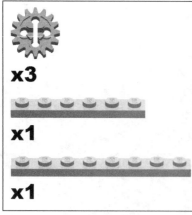

**x3**

**x1**

**x1**

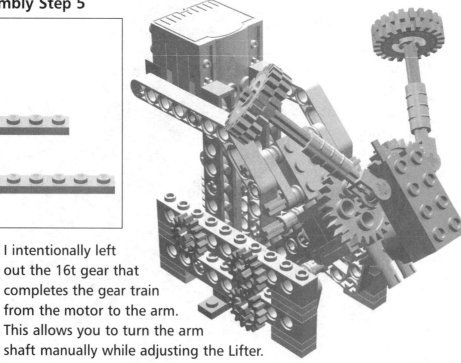

I intentionally left
out the 16t gear that
completes the gear train
from the motor to the arm.
This allows you to turn the arm
shaft manually while adjusting the Lifter.

## Adjusting the Lifter

ULK's arm works only if the gear mesh between the Arm shaft and Lifter shaft is adjusted properly. When the fingers are closed, and ULK's arm is raised to the full upright position, the 3x3 bent liftarms in the lifter should almost touch the 1x8 plate attached to the bottom of the Arm Assembly sub-assembly. If the arm cannot be raised to the full upright position, pull out the lifter shaft gear, rotate it by one tooth counter-clockwise relative to the arm shaft gear, and push it back in place.

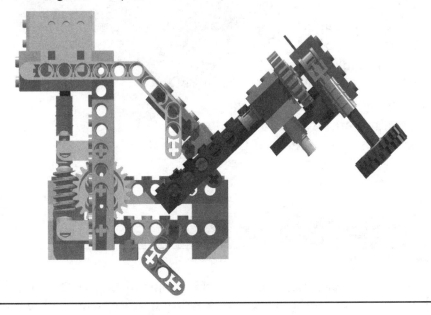

## Arm Assembly Step 6

The 16t gear installed here is an idler gear. An idler gear does not affect the gear ratio of a gear train. Idler gears are quite commonly used in machines to connect distant axles. They are also used to change the direction of rotation of the output shaft.

## Bricks & Chips...

## Gear Trains

A multi-stage gear train amplifies the torque of the LEGO motor to where it is sufficient to raise ULK's arm. A worm gear attached to the motor shaft turns a 24t gear. The 24t gear is attached to the same shaft as a 16t gear. The resulting gear reduction can be calculated by multiplying together the gear reduction for each stage.

- **Stage 1** (Worm gear to 24t gear) gear ratio = 24:1
- **Stage 2** (24t gear to 16t gear) gear ratio = 3:2
- **Overall** Gear ratio = 24 x 3:1 x 2, which equals 72:2 or 36:1

Wow! That's a lot of gear reduction.

# The Caster

ULK is a differential-drive robot. It has two main wheels that are used for locomotion and steering, and a third wheel that provides balance. The third wheel is a swivel caster that can pivot in its mount to point in the direction that ULK is arming.

## Caster Step 0

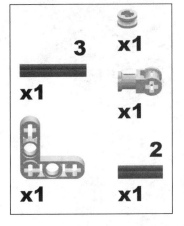

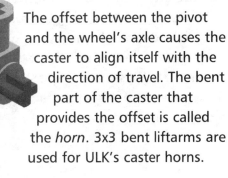

The offset between the pivot and the wheel's axle causes the caster to align itself with the direction of travel. The bent part of the caster that provides the offset is called the *horn*. 3x3 bent liftarms are used for ULK's caster horns.

## Bricks & Chips...

### Swivel Casters

The amount of offset provided by the horn has a big effect on the performance of a swivel caster. Casters with a large offset will track better and result in a more stable robot. Using casters with a small offset will make a robot's steering more responsive. If the offset is zero, the caster will not track, and the robot's motion is unpredictable.

### Caster Step 1

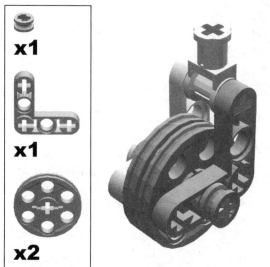

Casters have a tendency to steer a robot as they adjust to a new arming. Anyone who has pushed around a heavy shopping cart is familiar with this phenomenon. ULK uses a hard plastic wheel in its caster to minimize traction and reduce caster steer.

### Caster Step 2

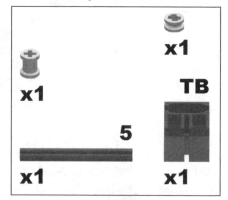

## Caster Step 3

x4

4
x1

10
x1

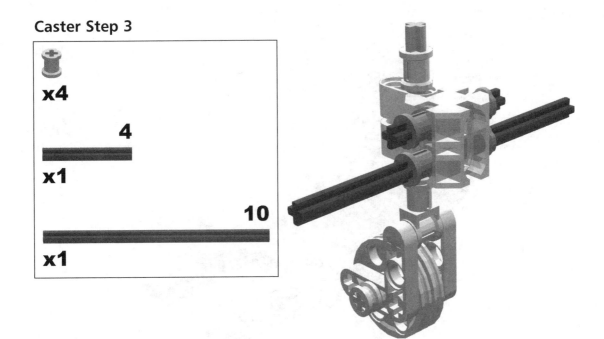

## Caster Step 4

x4

x2

10
x1

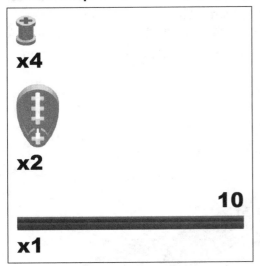

ULK's first caster mount had the #10 axles spaced one stud apart, but the weight of the robot caused the axles to bend, tilting the caster pivot slightly. To solve for this, the cams allow the mounting axles to be spaced two studs apart, resulting in less bending force applied to each axle. The improved caster has almost no tilt and tracks significantly better.

# The Base

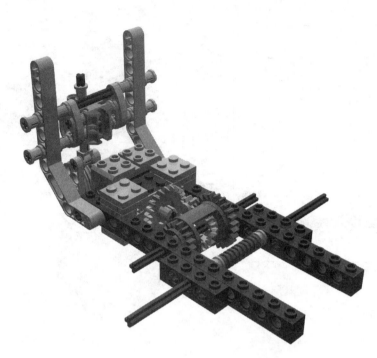

ULK uses just one motor for both locomotion and steering. When the motor spins counter-clockwise (forward direction according to the RCX), both wheels turn forward. When the motor spins clockwise (reverse direction), the right wheel spins backwards, while the left wheel is locked in place by a ratchet. This mechanism is called a ratchet splitter, and is useful when a motor is performing more than one operation.

## Base Step 0

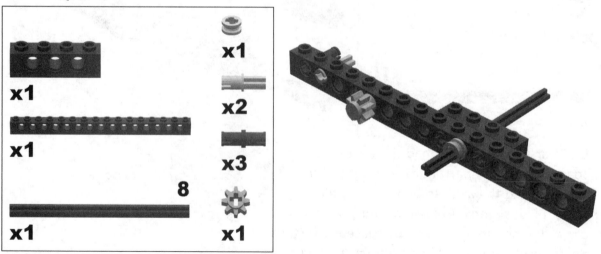

## Base Step 1a

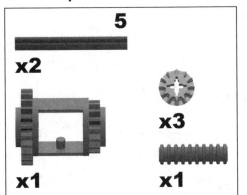

A differential drive won't travel in a straight line without some sort of tracking correction. Some robots use rotation sensors and special software to ensure that both wheels travel the same distance. ULK uses a clutch to coerce the left and right wheels into turning at the same speed.

A ribbed hose slipped over the drive axles makes a simple and effective clutch. I cut a piece three studs in length from the long teal hose. If you are nervous about defacing any of your LEGO pieces, Base Step 1b shows an alternate wheel attachment that doesn't require any cutting.

## Base Step 1b

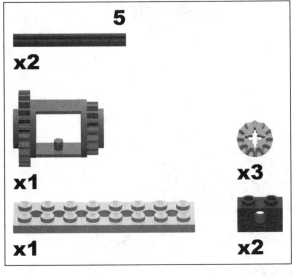

If you choose to not use the clutch, additional support for the drive axles help ULK drive a little straighter.

## Base Step 2

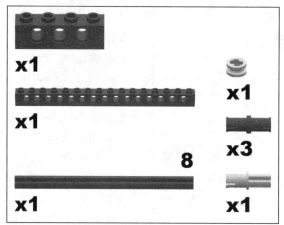

x1

x1

8

x1

x1

x3

x1

A key component
of the ratchet
splitter is the
differential
gear. The differential
gear distributes the torque
of the motor evenly
between the left and right
axles, allowing them to rotate
at different speeds

## Base Step 3

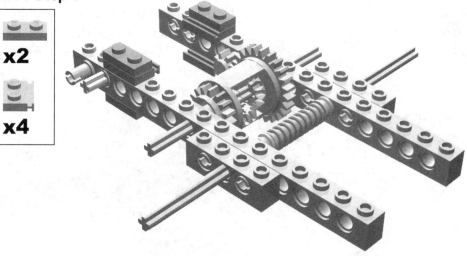

x2

x4

## Base Step 4

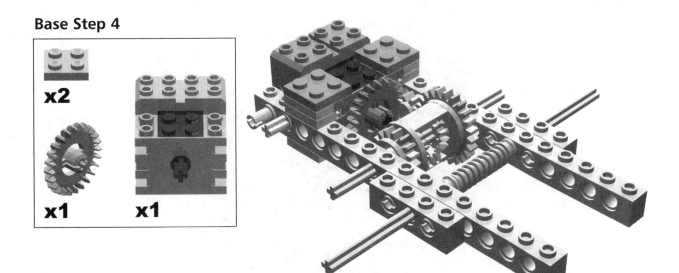

## Base Step 5

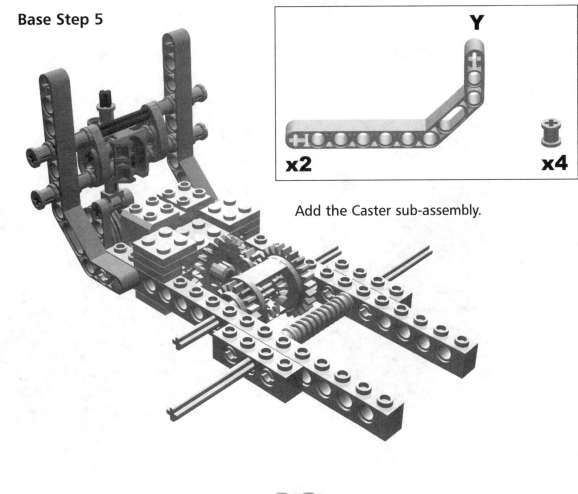

Add the Caster sub-assembly.

# Putting It All Together

ULK goes together very quickly now that all the major components are assembled.

## Final Step 0

Snap the Arm Assembly and
Base sub-assemblies together.

## Final Step 1

x2

x3

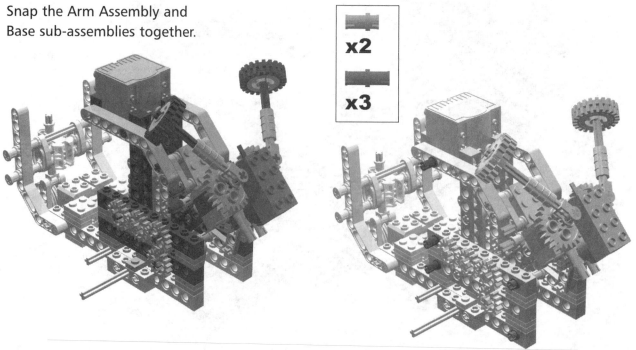

## Final Step 2

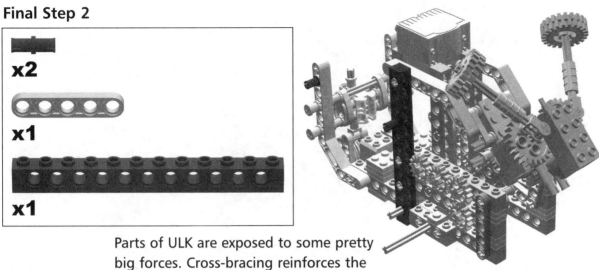

**x2**

**x1**

**x1**

Parts of ULK are exposed to some pretty big forces. Cross-bracing reinforces the snap-on connections, preventing them from coming apart. Without cross-bracing, ULK would self-destruct in a matter of seconds!

## Final Step 3

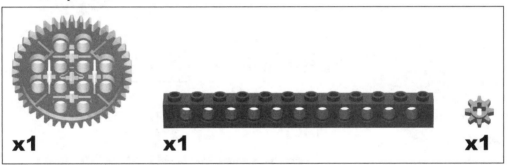

**x1**

**x1**

**x1**

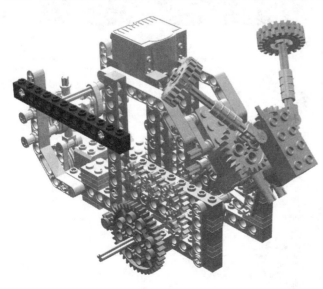

Additional bracing reinforces the caster mount. The 1x12 TECHNIC beam is also used to brace the RCX.

## Final Step 4

x2

x2

3

x1

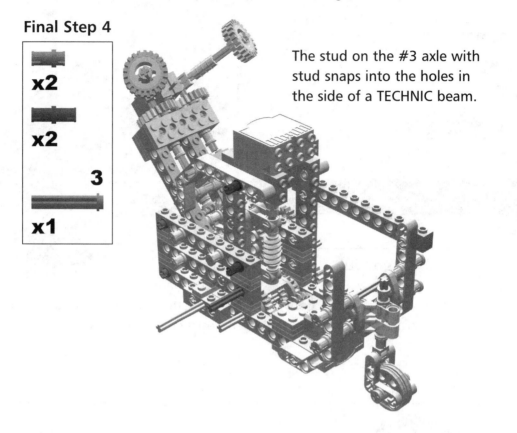

The stud on the #3 axle with stud snaps into the holes in the side of a TECHNIC beam.

## Final Step 5

x1

x1

x1

x2

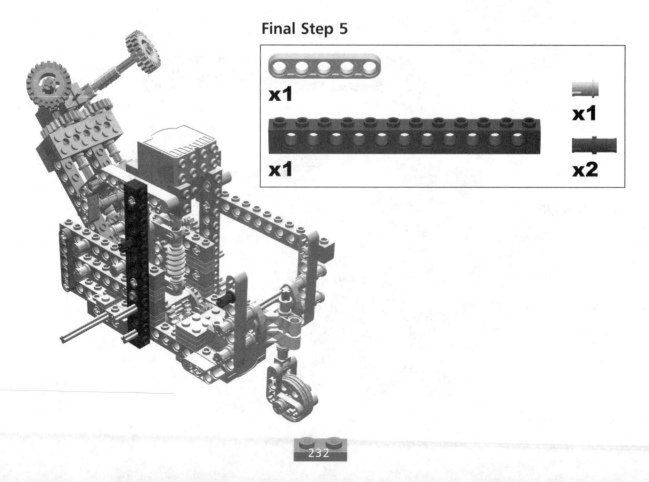

## Final Step 6

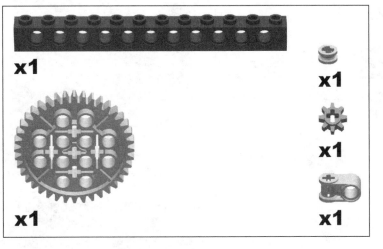

x1

x1

x1

x1

x1

The perpendicular axle joiner rests against the 40t gear and creates a ratchet. When the gear rotates counter-clockwise, the axle joiner skips across the top of the gear teeth. If the gear tries to rotate clockwise, the axle joiner catches in the gear teeth, preventing the gear from turning.

## Final Step 7

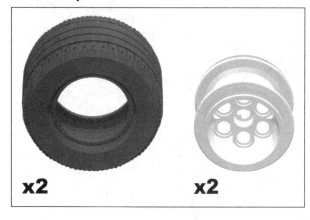

**x2**    **x2**

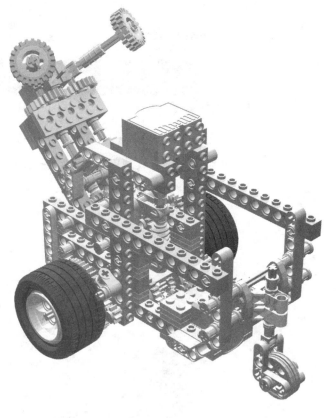

## Final Step 8

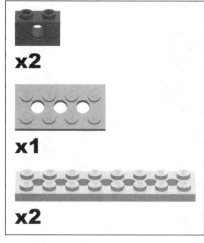

**x2**

**x1**

**x2**

These plates are used to mount the RCX.

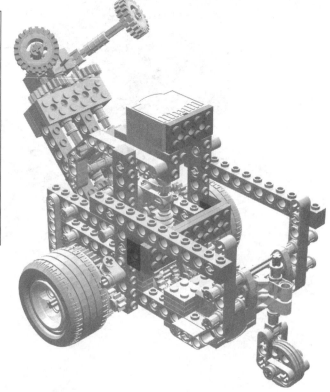

## Final Step 9

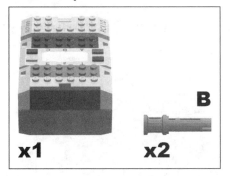

The 3L pins with stop bushings hold the RCX in place against the mounting plates. To change batteries, simply pull out the pins, unsnap the connector wires, and remove the RCX.

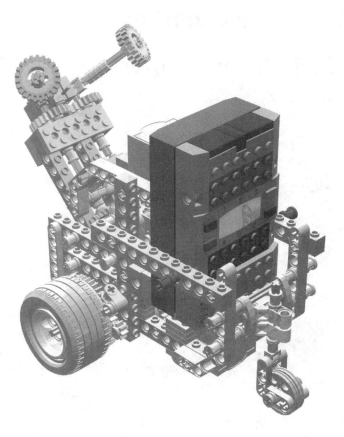

# Wiring Instructions

## Wiring the Arm Motor

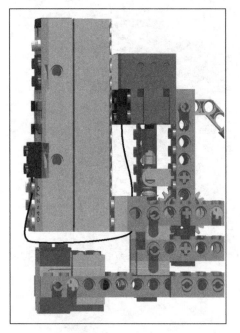

The motor for the Arm is attached to Output Port C. Route the 13-cm wire (the short ones) between the worm gear and RCX mounting plates, and around the bottom of the RCX.

## Wiring the Drive Motor

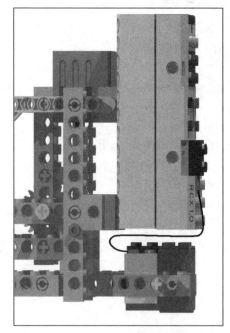

The drive motor is attached to Output Port A. Use a 13-cm connector wire.

## Wiring the Light Sensor

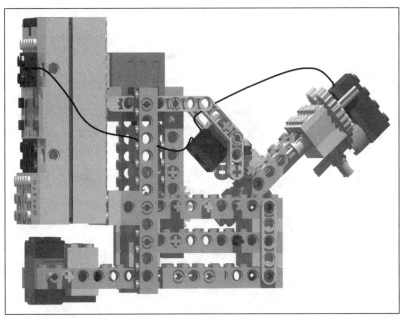

The light sensor is attached to Input Port 3. The wire attached to the sensor is too short to reach the RCX, so I used a 13-cm wire as an extension cord. Route the light sensor wire over the touch sensor, and snap the connector to the bottom of the light sensor. Attach the extension wire to the bottom of the light sensor connector, and route it around the arm motor mount, behind the 12L vertical cross brace beam, and around the side of the RCX.

# Wiring the Limit Switch

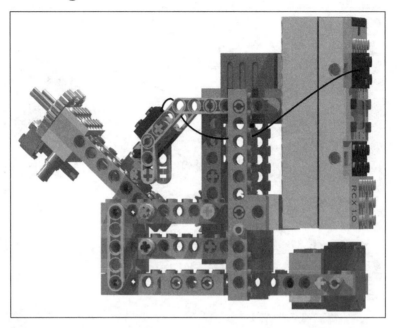

Use a 13-cm wire to attach the touch sensor to Input Port 1.
Route the wire under the touch sensor support liftarms,
around the arm motor mount, behind the 12L vertical cross
brace beam, and around the side of the RCX.

# Robot 9

## The SpinnerBot

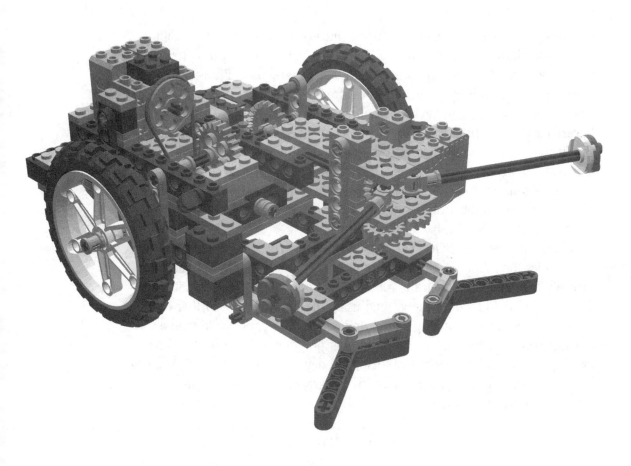

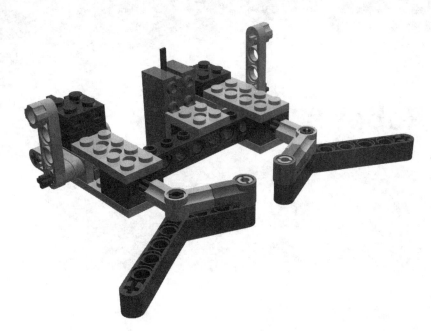

As a designer, you'll always be striving to squeeze a few more features into your robot without using any more RCX outputs. Using only two motors, the SpinnerBot allows you to experiment with driving, grabbing, lifting, and replacing objects. This robot gets its name from the fact that it spends a lot of time spinning: In order to navigate, it either moves forward or turns to the left. Nevertheless, the SpinnerBot can get wherever it needs to go using these two abilities, and when it gets there, it can grab and lift objects using the gripper arm powered by the second motor. The SpinnerBot uses touch sensors to detect the flags it collects, and it uses a light sensor in order to stay within the bounds of the dark oval on the Test Pad (the Test Pad is the large printed sheet that came with your RIS 2.0 kit).

Using the language NQC (Not Quite C), I have written a Capture-the-Flag program for the robot, which hunts around the confines of the Test Pad for a specially designed flag. When the robot encounters a flag, it captures it using the grabbers, and lifts it off the ground using the lift arm. As a further challenge, you might try setting up your own arena with a dark border and a medium-shaded "home" area. You could then program SpinnerBot to search for flags outside its home, retrieve them, and deposit them in its nest. You can find the Capture-the-Flag program on the Syngress Solutions Web site (www.syngress.com/solutions).

# The Left Side Frame

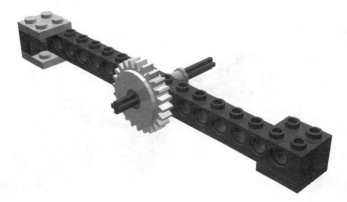

The Left Side Frame sub-assembly will hold the left drive wheel in the finished robot, and provides structural rigidity to the frame.

### Left Side Frame Step 0

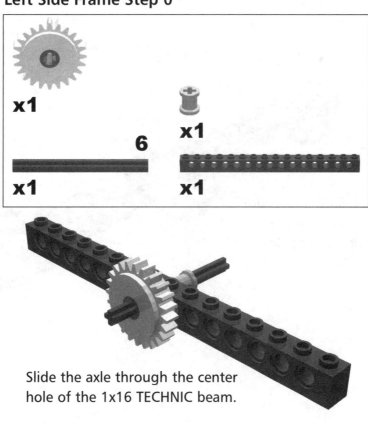

Slide the axle through the center hole of the 1x16 TECHNIC beam.

## Left Side Frame Step 1

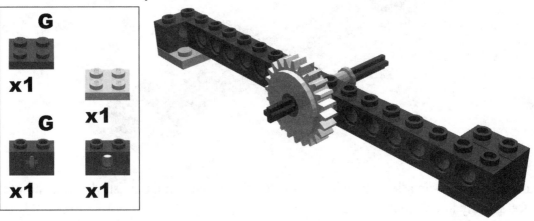

**G**

x1

x1

**G**

x1

x1

## Left Side Frame Step 2

x1

Set the Left Side Frame sub-
assembly aside. You will
need it when you get to
**Final Step 10**.

# The Right Side Frame

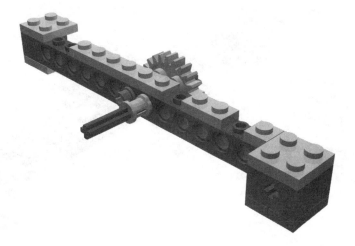

The Right Side Frame sub-assembly is the complement to the Left Side Frame sub-assembly, but uses a different gear. The right wheel will be able to move forward, but not backward.

### Right Side Frame Step 0

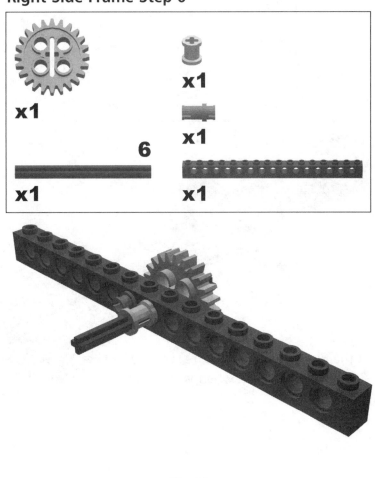

## Right Side Frame Step 1

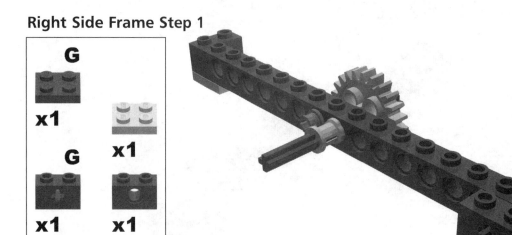

## Right Side Frame Step 2

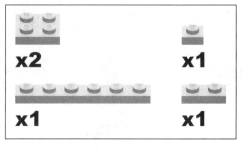

Set the Right Side Frame sub-assembly aside. You will need it when you get to **Final Step 10**.

# The Outer RCX Bracket

The Outer RCX Bracket sub-assembly uses the light gray pins to attach it to the frame. These pins come apart much more easily than the black pins with friction, and allow you to easily remove the RCX.

### Outer Bracket Step 0

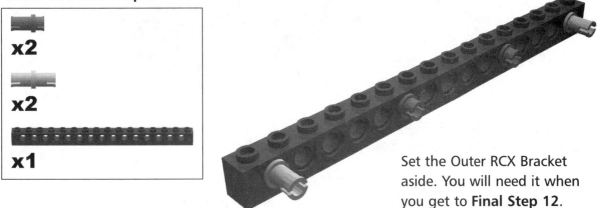

Set the Outer RCX Bracket aside. You will need it when you get to **Final Step 12**.

# The Inner RCX Bracket

The Inner RCX Bracket sub-assembly and the Outer RCX Bracket sub-assembly will be used later to securely hold the RCX within the body of the robot.

### Inner RCX Bracket Step 0

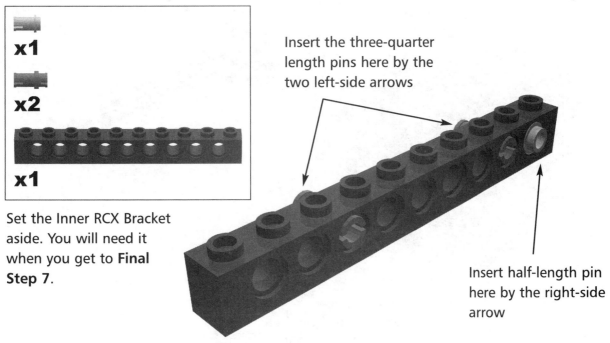

Insert the three-quarter length pins here by the two left-side arrows

Insert half-length pin here by the right-side arrow

Set the Inner RCX Bracket aside. You will need it when you get to **Final Step 7**.

# The Driving Mechanism

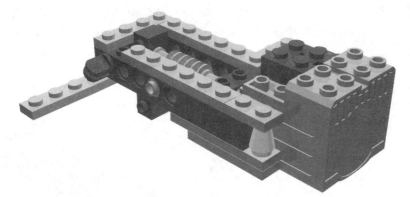

The Driving Mechanism sub-assembly will provide power to the wheels of the SpinnerBot. It uses a worm gear to give a powerful 24:1 speed reduction.

**Drive Step 0**

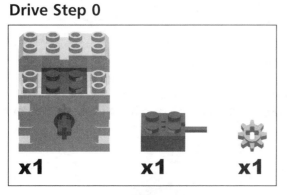

**Drive Step 1**

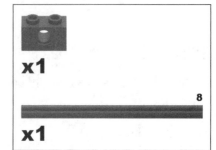

Set the motor off to the side. You will need this part when you get to **Drive Step 10**.

## Drive Step 2

**x1**

**x1**

Hang the 1x4 TECHNIC beam from the drive axle.

## Drive Step 3

**x1**

**x1**

## Drive Step 4

**x1**

**x1**

## Drive Step 5

**x1**

**x1**

**x1**

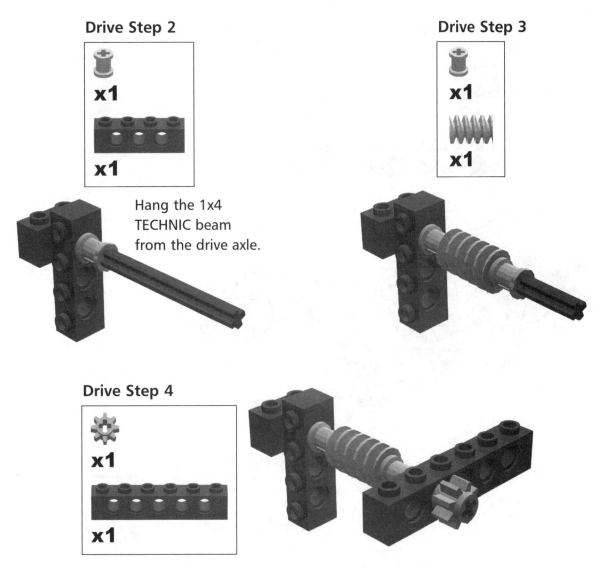

The black TECHNIC pin used in this step is normally used to attach decorative claws to your robots. However, it turns out that LEGO electrical wires fit tightly between the exposed prongs of this piece. You're going to use it later on to secure the wire from the light sensor.

## Drive Step 6

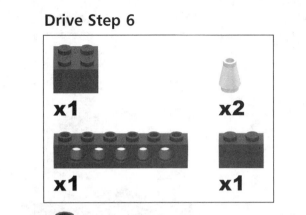

x1  x2

x1  x1

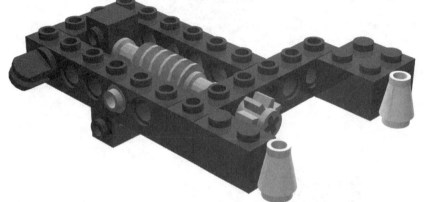

## Drive Step 7

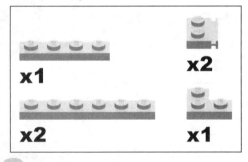

x1  x2

x2  x1

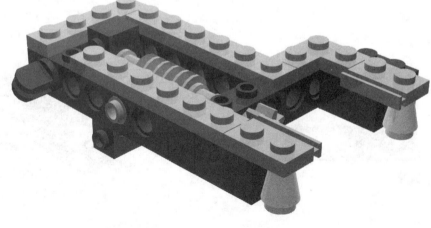

## Drive Step 8

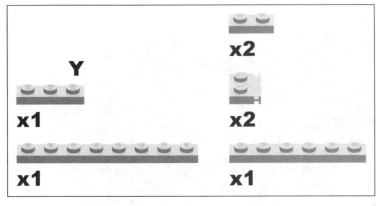

**Y**

x1

x2

x2

x1

x1

Flip the Drive Mechanism sub-assembly over to attach the plates to the underside.

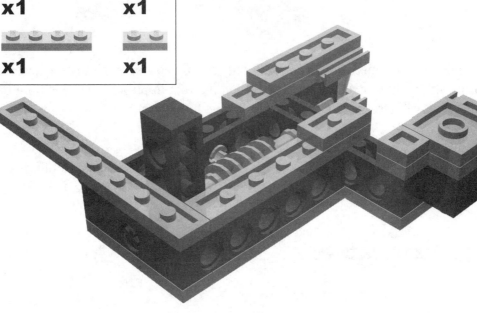

## Drive Step 9

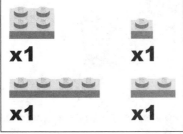

x1

x1

x1

x1

## Drive Step 10

Locate and attach the assembly created in **Drive Step 0**.

Set the Driving Mechanism sub-assembly off to the side. You will need this part when you get to **Final Step 14**.

# The Lower Arm Drive

You will assemble the drive mechanism for the lifting arm in two stages. This is the lower half of the mechanism, which will give a 24:1 speed reduction in order to prevent the flags from being tossed into the air. It will also serve as an armrest for the lifting arm.

## Lower Arm Drive Step 0

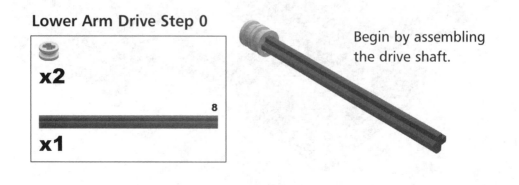

x2

8

x1

Begin by assembling the drive shaft.

### Lower Arm Drive Step 1

**x1**

**x1**

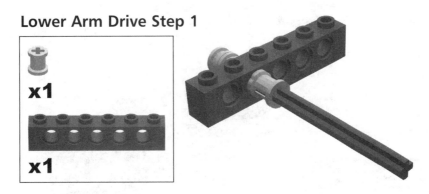

### Lower Arm Drive Step 2

**x1**

**x1**

### Lower Arm Drive Step 3

**x1**

**x1**   **x1**

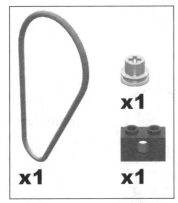

Slip a blue rubber band over the pulley piece. You will attach it to the arm motor later, in **Final Step 16**.

Place this part off to the side. You will need it later on in **Lower Arm Drive Step 6**.

## Lower Arm Drive Step 4

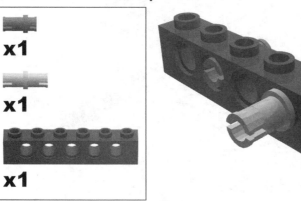

## Lower Arm Drive Step 5

x1

This axle connector is used as a ratchet to prevent the right wheel from rotating backward.

## Lower Arm Drive Step 6

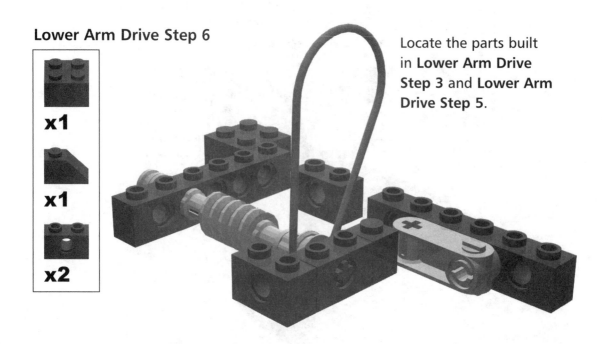

x1

x1

x2

Locate the parts built in **Lower Arm Drive Step 3** and **Lower Arm Drive Step 5**.

## Lower Arm Drive Step 7

## Lower Arm Drive Step 8

The smooth-topped plate is the lower half of a hinge; pull apart the hinge, as you will only be using the lower half. The lifting arm will rest on the smooth surface.

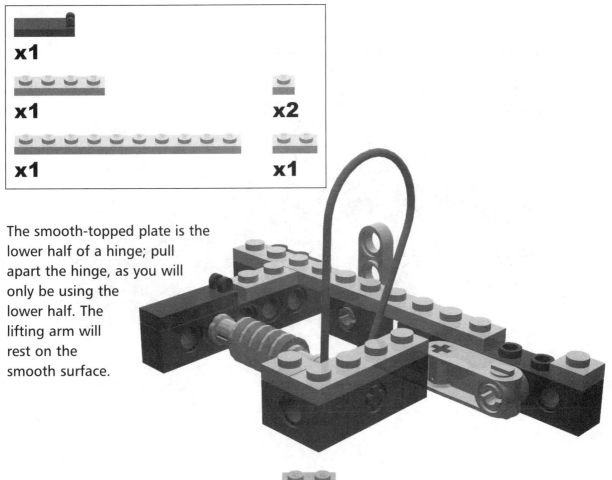

## Lower Arm Drive Step 9

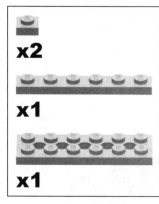

x2

x1

x1

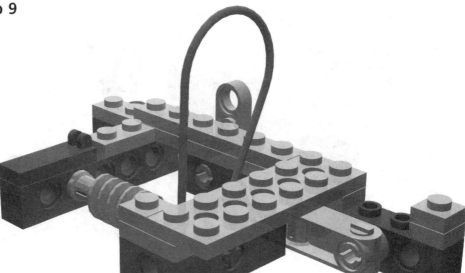

## Lower Arm Drive Step 10

x1

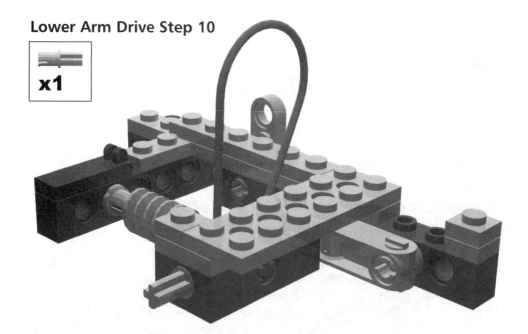

## Lower Arm Drive Step 11

Attach the yellow, sloped brick by sliding it over the fingers of the black hinge plate and securing it with the yellow 1x3 plate. This sloped brick serves as a rest for the lifting arm.

## Lower Arm Drive Step 12

Flip over the Lower Arm Drive sub-assembly to attach plates to the underside.

Set this sub-assembly off to the side. You will use this piece in **Final Step 15**.

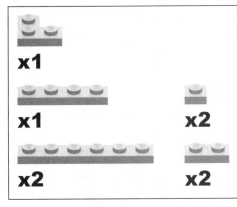

# The Arm Drive

This is the upper half of the power train for the lifting arm of the SpinnerBot. The motor here will drive a pulley belt to the Lower Arm Drive sub-assembly. The size of the pulley on the motor is larger, giving about a 3:1 speed increase, which will be offset by the worm gear in the Lower Arm Drive sub-assembly.

### Arm Drive Step 0

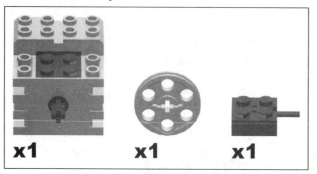

**x1**     **x1**     **x1**

Connect the pieces to the motor as shown. Place this assembly off to the side. You will need this in **Arm Drive Step 6**.

## Arm Drive Step 1

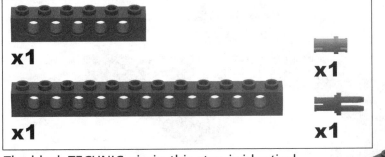

The black TECHNIC pin in this step is identical to the one used earlier in **Drive Step 5**. You will use it later to secure the wire from the touch sensors to the side of the robot.

## Arm Drive Step 2

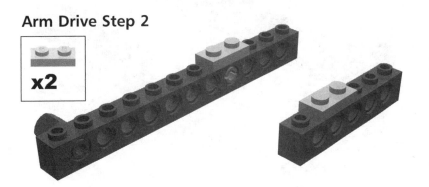

## Arm Drive Step 3

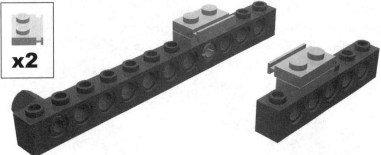

### Arm Drive Step 4

x1

x1

### Arm Drive Step 5

x1

### Arm Drive Step 6

Locate the motor built in **Arm Drive Step 0**, and place it in the motor holder you created using the two TECHNIC bricks.

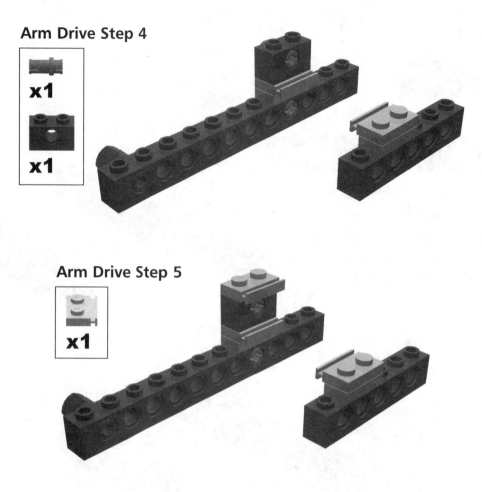

## Arm Drive Step 7

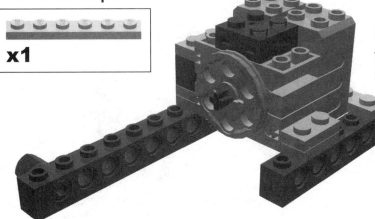

**x1**

Place the 1x6 plate behind the motor to prevent it from sliding out of its holder.

## Arm Drive Step 8

**x1**

Secure the black 1x7 TECHNIC half-beam to the two three-quarter length pins. When you assemble the entire robot, this will line up with two more three-quarter length pins, providing a secure reinforcement to the model.

Slide the Arm Drive sub-assembly off to the side; you'll need it in **Final Step 16**.

# The Left Gripper

Now you will start to assemble the business end of the lifting arm. The Left Gripper moves in concert with the Right Gripper to grab a flag.

## Left Gripper Step 0

x1

4

x1

## Left Gripper Step 1

x1

x1

x1

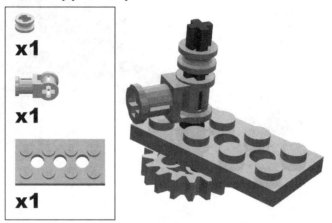

Leave a small gap between the half-bushing and the axle joiner. The half-bushing will prevent the gripper shaft from rattling up and down once the gripper is complete.

## Left Gripper Step 2

x1

B

x1

10

x1

The 2x2 round plates add a touch of color, and also serve the function of preventing the flag from slipping off the end of the grippers.

Set the Left Gripper sub-assembly aside; you will need it in **Spinner Arms Step 0**.

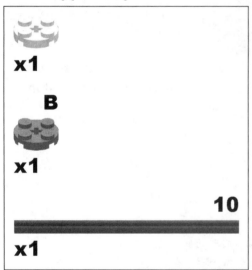

# The Right Gripper

The Right Gripper sub-assembly differs from the Left Gripper in that it contains a 12t bevel gear, which provides power to the pair of grippers. The 16t gear will drive the Left Gripper so that they move as a pair.

### Right Gripper Step 0

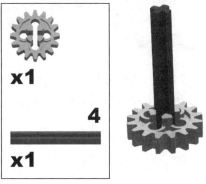

### Right Gripper Step 1

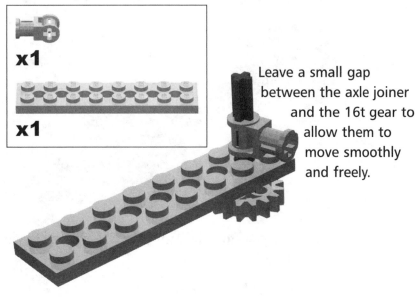

Leave a small gap between the axle joiner and the 16t gear to allow them to move smoothly and freely.

## Right Gripper Step 2

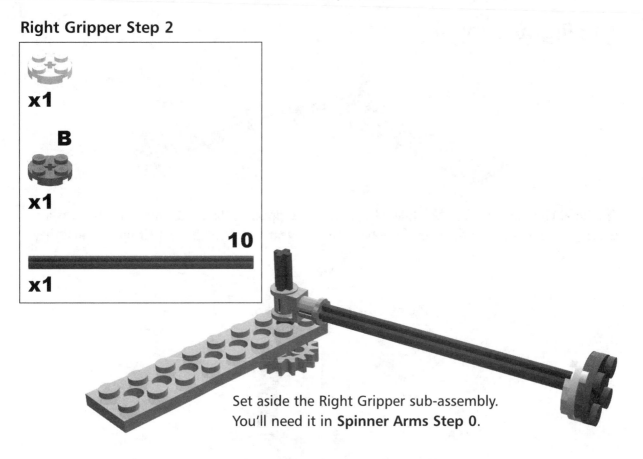

**x1**

**B**

**x1**

**10**

**x1**

Set aside the Right Gripper sub-assembly.
You'll need it in **Spinner Arms Step 0**.

# The Gripper Drive Shaft

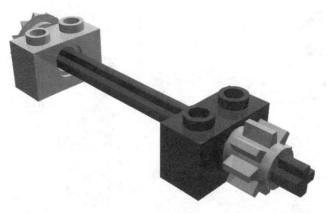

The Gripper Drive Shaft transfers power from the Arm Pivot (the next sub-assembly you'll build) to the Gripper Arms you have just built.

## Gripper Drive Shaft Step 0

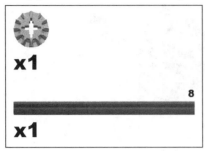

The 12t bevel gear goes on the very end of the axle. If you allow the axle to protrude, it may jam against the Right Gripper Arm.

## Gripper Drive Shaft Step 1

Slide the 8t gear onto the axle until there is a half-stud width of axle showing at the end. Achieve this spacing by sliding a spare half-length bushing onto the tip of the axle, snugging the gear up to it, and then removing it.

You will use this sub-assembly in **Spinner Arms Step 2**.

# The Arm Pivot

This sub-assembly serves two important functions: It provides power to the Gripper Arms, and once those arms are closed around an object, it serves as the pivot for the entire lifting arm. This happens because the 8t gear in the Gripper Drive Shaft can no longer turn once the grippers are closed. The force from the 24t crown gear in the Arm Pivot lifts the entire arm instead, and the flag leaves the ground.

### Arm Pivot Step 0

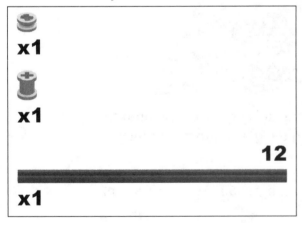

x1

x1

**12**

x1

Leave a one-and-a-half stud space at the end of the axle, and a one-stud space between the two bushings.

## Arm Pivot Step 1

Slide the gear up the axle until it sits snugly between the two half-length bushings.

## Arm Pivot Step 2

Leave a one-stud gap between the 24t crown gear and the first six-stud TECHNIC beam.

## Arm Pivot Step 3

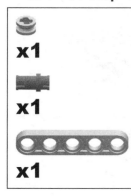

The three-quarter-length pin serves as a stop to limit how far the arm can lift. Later, you will attach the vertical piece to the main Drive Mechanism in order to provide reinforcement.

## Arm Pivot Step 4

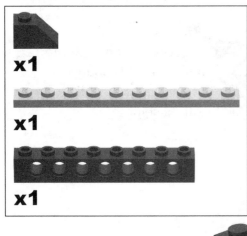

x1

x1

x1

Set the Arm Pivot aside for now. You will need it in **Spinner Arms Step 3**.

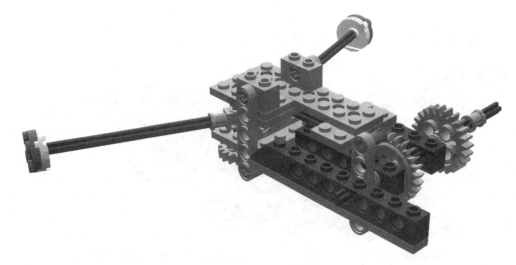

# The Spinner Arms

The Spinner Arms are the business end of the SpinnerBot. The power from the Arm Pivot axle sub-assembly will open and close the grippers, and will also raise and lower the Arm.

### Spinner Arms Step 0

Locate the Right and Left Gripper sub-assemblies built previously and place them beside each other, as shown.

### Spinner Arms Step 1

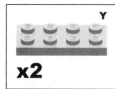

As you attach the two grippers using the yellow plate, try to match the positions of the grippers while the 16t gears mesh. There will always be a slight difference in orientation (the width of half a tooth, or 15 degrees), but if they are far off, the grippers will not close properly. Once you attach the plate, you can test rolling them open and closed in unison.

### Spinner Arms Step 2

Locate the Gripper Drive Shaft sub-assembly built previously and attach it as shown.

Make sure that the two 1x2 TECHNIC beams on the shaft are as far apart as possible before you snap them onto the 2x8 plate.

## Spinner Arms Step 3

Locate the Arm Pivot sub-assembly
built previously, and attach it as shown.

The 24t crown gear and 8t gear
should mesh neatly.

## Spinner Arms Step 4

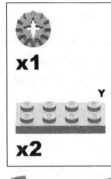

Slide the 12t gear, teeth-down, onto the axle
of the Right Gripper sub-assembly, until it meshes
with the gear on the Drive Shaft sub-assembly. You will
want to support the end of the #4 axle with a fingertip as
you do so, to prevent it slipping down from the pressure.

## Spinner Arms Step 5

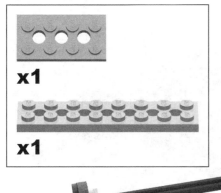

**x1**

**x1**

The 2x8 plate should fit snugly on top of the 12t bevel gear. The thin ring on the back of the gear will actually inset into the underside of the 2x8 plate.

## Spinner Arms Step 6

**Y**

**x2**

## Spinner Arms Step 7

**x2**

Each part required for this step is a 2x2 TECHNIC plate with an unusual underside: Underneath each plate is a rounded hole that allows an axle or TECHNIC pin to be inserted crosswise through the plate.

Flip the sub-assembly over to attach these plates to the underside. The 16t gears from the grippers will overlap the plates slightly. You may need to loosen the gears slightly on their axles to allow the plates above them. Test the grippers to make sure that they still open and close smoothly.

## Spinner Arms Step 8

**x4**

Turn the assembly upright again to snap the three-quarter length pins into each of the two plates and two bricks you just attached.

## Spinner Arms Step 9

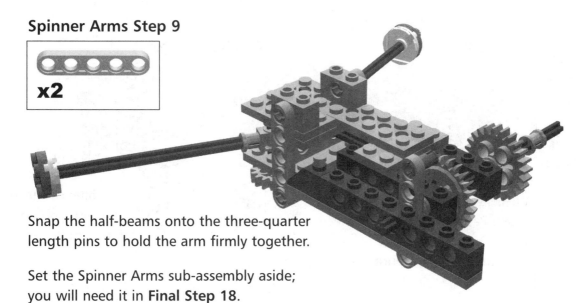

**x2**

Snap the half-beams onto the three-quarter length pins to hold the arm firmly together.

Set the Spinner Arms sub-assembly aside; you will need it in **Final Step 18**.

# The Touch Pads

Right Touch Pad

Left Touch Pad

The Touch Pads are the probes with which the SpinnerBot detects a flag. Each Touch Pad is built so that its center of gravity is below the connector. As a consequence, the pads will hang level with the ground regardless of how you orient the SpinnerBot. The Touch Pads are just wide enough to sense the flag if it is in range of the grippers, and to miss it otherwise.

## Touch Pads Step 0

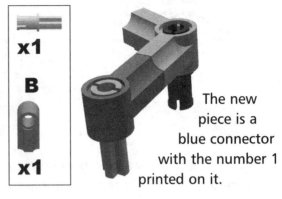

We'll start by building the Right Touch Pad first. The angled piece in this step is a gray angle connector with the number 5 printed on it.

## Touch Pads Step 1

The new piece is a blue connector with the number 1 printed on it.

## Touch Pads Step 2

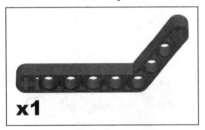

This is the completed Right Touch Pad. Set this off to the side. You will use this sub-assembly in **Touch Mechanism Step 3**.

## Touch Pads Step 3

Now we'll build the Left Touch Pad. Again, in this step, the angled piece is a gray angle connector with the number 5 printed on it.

## Touch Pads Step 4

The new piece is a blue connector with the number 1 printed on it.

**Touch Pads 5**

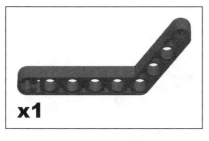

**x1**

This is the completed Left Touch Pad. Set this off to the side. You will use this sub-assembly in **Touch Mechanism Step 3**.

# The Touch Mechanism

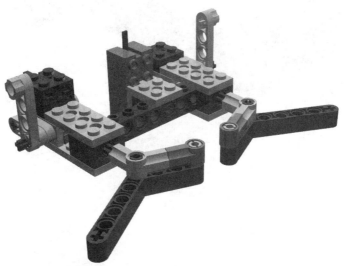

All of the sensors on the SpinnerBot are located on this sub-assembly. The touch sensors are plunger-style, and are wired in parallel. This means that the SpinnerBot does not normally distinguish between one sensor or the other–rather, they are both attached to the same Input Port, so touching either will trigger the flag-grabbing action.

However, if you wire each sensor to a separate port, you can enhance your program to give the SpinnerBot a more sophisticated behavior.

The light sensor is also located here, and as with any line-sensing robot, it is pointed directly at the ground.

The entire Touch Mechanism sub-assembly barely clears the ground. In fact, the Touch Pads serve as forward skids for the robot when it lifts a flag, since the heavy flag moves the center of gravity forward of the wheel axles.

## Touch Mechanism Step 0

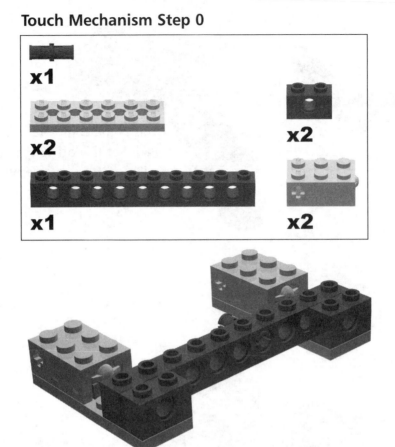

## Touch Mechanism Step 1

## Touch Mechanism Step 2

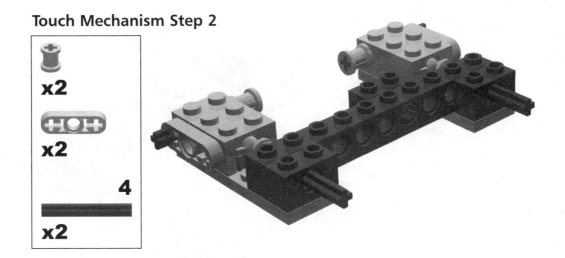

## Touch Mechanism Step 3

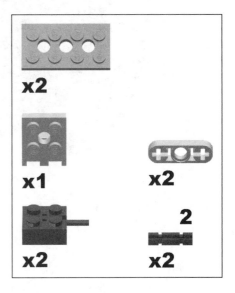

Locate and attach the Right and Left Touch Pads as shown.

The two wire ends depicted in the parts list both belong to a single five-inch (13 cm) wire. Connect one touch sensor to the other with the short wire.

## Touch Mechanism Step 4

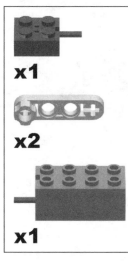

Set the Touch Mechanism sub-assembly aside. You will need it in **Final Step 20**.

The wire end in this step belongs to a 48-inch (120 cm) wire. Connect one end of this wire on top of the short wire on the right touch sensor, as shown.

# Putting It All Together

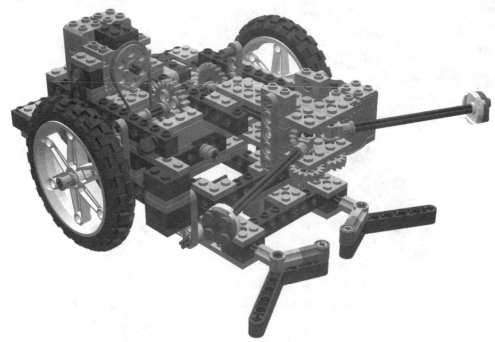

Now you will build the chassis of the SpinnerBot, and attach each of the sub-assemblies you've set aside. Take care to position each one properly.

## Final Step 0

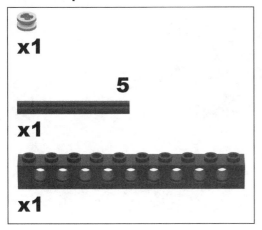

x1

**5**

x1

x1

Begin by assembling the transmission. Slide the half-length bushing onto the axle, leaving one-and-a-half studs of axle exposed on the near side of the bushing.

## Final Step 1

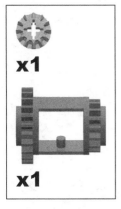

x1

x1

Place the 12t bevel gear on the short stub inside the barrel of the differential.

## Final Step 2

x1

x1

As you slide the differential housing onto the axle, slip the 12t gear onto the end of the axle. This will hold the loose 12t gear from the previous step in place.

The new 12t gear should just barely cover the end of the axle.

## Final Step 3

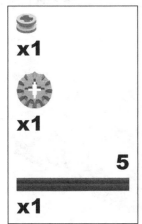

x1

x1

**5**

x1

Add the second axle, sliding the 12t gear over the end as it enters the differential. Continue sliding the axle through the gear until a full stud's worth of the end of the axle is exposed inside the differential. It should just brush the axle tip on the opposite side of the differential.

### Final Step 4

x1

x1

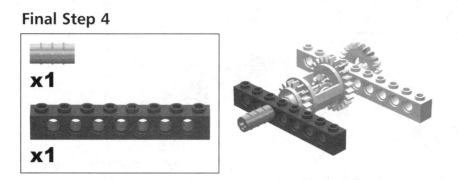

### Final Step 5

x2

6

x1

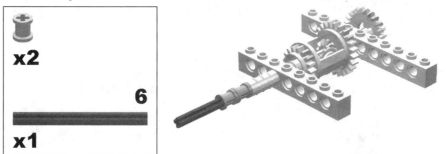

### Final Step 6

x1

x1

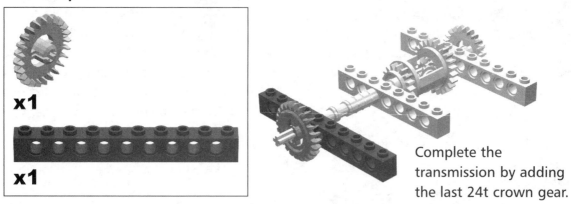

Complete the transmission by adding the last 24t crown gear.

### Final Step 7

x1

x3

You will need the Inner RCX Bracket sub-assembly that you set aside earlier. Make certain that it has all three pins—it would be very difficult to add them later on!

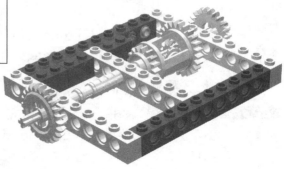

## Final Step 8

**x1**

**x1**

**x1**

**x3**

Reinforce the chassis with a cap of plates.

## Final Step 9

**x1**

**x1**

**x1**

**x1**

**x2**

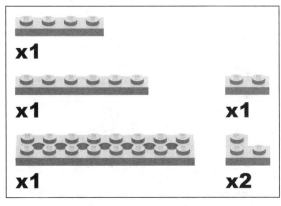

Now flip over the chassis and add plates to the underside. This makes a rigid frame for the rest of the SpinnerBot.

## Final Step 10

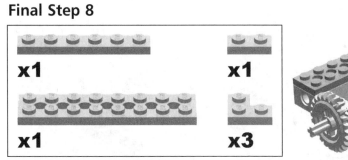

Attach the Left and Right Side Frame sub-assemblies to the SpinnerBot, as shown. Attach the Left Side Frame first by sliding the exposed end of the Left Side Frame axle into the hole on the end of the chassis. Then repeat the process with the Right Side Frame.

## Final Step 11

**x1**

The RCX snaps onto the three-quarter length pins on the Inner RCX Bracket sub-assembly. The position of the RCX is key to a mobile robot, as it is typically the most important factor in the balance of the robot.

## Final Step 12

Snap the Outer RCX Bracket sub-assembly onto the end of the chassis, securing the RCX. The Outer RCX Bracket and RCX can be readily removed.

## Final Step 13

**x1**

**x1**

Flip the robot over and add the skid plate to the underside of the RCX. This plate provides a reasonably low-friction bearing for the SpinnerBot, since it is almost perfectly balanced over the wheels.

## Final Step 14

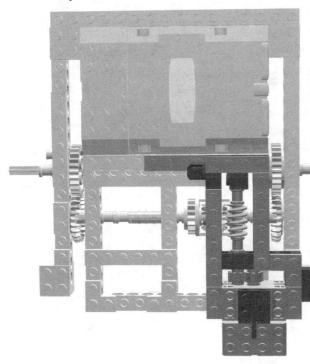

Attach the Drive Mechanism sub-assembly as shown. The dangling 1x4 TECHNIC beam from the Drive Mechanism should slip between the differential and the Inner RCX Bracket, aligning with the half-length pin there. Once you have firmly attached the sub-assembly, reach in with your fingers or a narrow plate and firmly snap the 1x4 TECHNIC beam onto the half-length pin. This will prevent the Drive Mechanism from gradually working apart.

## Final Step 15

Attach the Lower Arm Drive sub-assembly as shown. As you attach the drive, lift the dangling axle connector you added in **Lower Arm Drive Step 5** and set it on top of the 24t gear from the Right Side Frame. This will prevent the right wheel from rotating backward. Test this by rotating the protruding right axle either way. It should move freely in one direction and lock in the other. If it does not, try re-seating the sub-assembly.

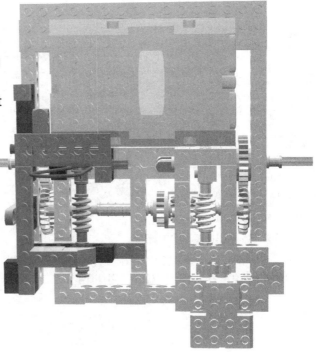

## Final Step 16

Attach the Arm Drive sub-assembly as shown. The dangling black half-beam should be aligned with the two protruding pins further down on the body of the SpinnerBot. Once the Arm Drive is seated, this half-beam has to be snapped onto the pins, securing the entire column in place.

Take the blue rubber band that is loosely attached to the Lower Arm Drive's shaft, and slip it securely around the groove of the small pulley on the shaft. Once it is in place, stretch the band around the larger pulley on the arm motor. The band be lined up vertically, and it should fit snugly without crossing over itself.

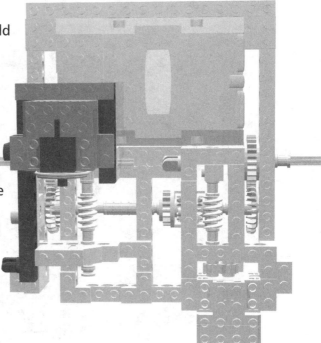

## Final Step 17

Insert pins here

Looking at the left side of the Arm Drive, connect the two liftarms to the side of the motor housing using the axle pins. This will prevent the housing from coming loose under strain.

**Final Step 18**

Attach the Spinner Arms sub-assembly. Slide the loose end of the Arm Pivot into the right side of the robot. The end should be secured using the 5-length gray half-beam from the Lower Arm Drive. Firmly press the beam at the other end of the Arm Pivot onto the Drive Mechanism. The 5-length gray half-beam at this end of the

Arm Pivot
will align with and clip onto a three-quarter length pin attached to the side of the Drive Mechanism.

**Final Step 19**

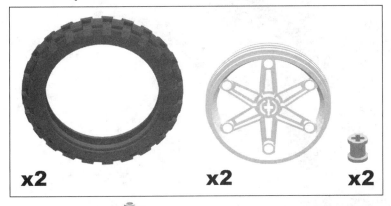

**x2**          **x2**          **x2**

Attach the wheels and tires. Take a moment to rotate each wheel. The left wheel should freely spin backward while the right wheel freely spins forward. If the wheels do not move freely, take a moment to adjust any parts that might be in contact with them.

## Final Step 20

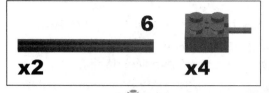

Secure the Touch Mechanism by pinning each upright arm with a #6 length axle inserted into the green 1x2 brick with axle hole.

- Connect the light sensor to Input Port 1 (I ran the wire through the body and secured it using the pin on the corner of the Drive Mechanism).

- Connect the touch sensors to Input Port 2 (I ran a long wire down the right side of the robot, securing the excess with a black rubber band).

- Connect the arm motor to Output Port A.

- Connect the main drive motor to Output Port C.

# The Spinner Flag

These are the flags that the SpinnerBot will pick up as it travels around its track. You can build two of these with the extra parts in the RIS.

These flags are based on a design that rtlToronto (among others) used in a competition in which the robot had to pick up the flag to win. The large top and the skinny body make the flag easy to pick up.

## Spinner Flag Step 0

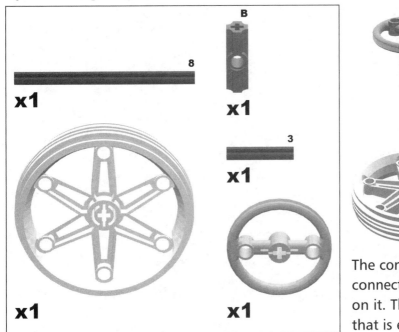

The connector is a blue angle connector with the number 2 printed on it. The large pulley provides a stop that is easy for the SpinnerBot to grab.

## Spinner Flag Step 1

**x1**

**x1**

The rubber tire adds a lot of weight to the flag, and makes the bottom stickier. Both of these properties make the flag easier to detect with the touch sensors. Remember you can build two of these!

# Troubleshooting

Upon running the Capture-the-Flag program, the SpinnerBot should try to lower its lifting arm and open the grippers. If the arm goes up instead of down, try reversing the orientation of the electrical connector on the arm motor. If the robot begins moving without trying to reset the arm, you need to check that the arm motor is connected to RCX Output Port A, and that the main drive motor is connected to Output Port C.

Once the arm has been reset, SpinnerBot will begin trying to move forward. If it moves forward but turns while doing so, one of the wheels is having a hard time turning. Make sure no loose wires or pieces are touching the wheel, and try turning the wheel by hand to ensure it spins freely. You may need to reach into the underside of the robot and wiggle the gears of the drive shaft to make sure they are loose enough to not bind on a beam. If SpinnerBot moves backward and turns instead of moving forward, check to see that the light sensor is in place and is looking at a white part of the Test Pad. If so, you will need to reverse the electrical connector on the main drive motor.

The SpinnerBot should reverse and turn when it reaches the black oval. If it fails to change direction at all, check that the light sensor is in place and connected to the RCX. By pressing the **View** button on the RCX, you can see what light level the light sensor is reporting. Under very bright conditions such as direct sunlight, you may need to lower the *WHITE* threshold in the program to a lower value. If the light sensor is working properly, but the SpinnerBot reverses without turning, the ratchet on the right wheel is not engaging. This ratchet was added in **Lower Arm Drive Step 5**, and should have been set on top of the 24t gear in **Final Step 15**. You may need to remove the Arm Drive sub-assembly and try repositioning it according to the instructions. You can tell if the ratchet is working: The right wheel should turn forward freely with a slight clicking sound, but should not turn backward at all.

If the SpinnerBot does not detect a flag upon bumping it, test the bumper by pressing it while watching for the small arrow to appear under Input Port 2 on the RCX LCD display. If no arrow appears when you press the bumper, you will need to check the wires connecting the touch sensors to Input Port 2. If the arm motor turns on and the grippers close but the arm does not move, the rubber band might not be seated properly on its pulleys. Make sure the blue band travels around the groove in the smaller pulley on the Lower Arm Drive sub-assembly, and around the groove in the larger pulley on the motor of the Upper Arm Drive sub-assembly.

# Playing Capture-the-Flag

The Capture-the-Flag program causes the SpinnerBot to roam around the inside of the black oval on the Test Pad, searching for a flag to grab. Try placing a flag on the middle of one of the straight sides of the oval, and put the SpinnerBot somewhere within the oval.

When you run the program, the SpinnerBot will begin by lowering its lift arm and opening the grippers. (If this doesn't happen, see the earlier Troubleshooting section.) Now the robot is ready to grab any flag it encounters. SpinnerBot will then begin traveling straight ahead until it encounters a black line or bumps a flag. When it encounters a black line, it should back up while turning right until it clears the line, and then resume its advance. Upon bumping the flag, SpinnerBot will stop moving and will activate its grippers. The grippers will close on the shaft of the flag, grabbing it firmly. Once the grippers are closed, the arm will begin to lift the flag over the top of the robot. After lifting the flag, the Spinner will shut down, having completed its task.

## Advanced Challenges

The SpinnerBot is a capable platform for programming advanced robot behaviors. Try out some of the following tasks to test your robotics mettle:

- After the robot captures a flag, have it play a sound, and set the flag down again. Have it back away from the flag, drive off in another direction, and then continue searching for other flags.

- Change the behavior of the robot when it encounters a black boundary line. By executing a larger turn, or a random-distance turn, your robot will zig-zag across the Test Pad rather than following the line closely.

- Wire each of the touch sensors separately so that the SpinnerBot can tell which direction an obstacle is in.

- Add an additional sensor from another LEGO set, and use it for more complex behaviors, such as detecting the edge of a table or determining whether the grippers missed the flag.

- Wire the touch sensors to the same input as the light sensor, and treat extremely high light readings as touch-sensor events. This frees another sensor input, so now SpinnerBot can detect a table edge *and* tell whether it has successfully grabbed a flag.

- Use black tape to outline your own roaming area.

- Place a sheet of medium-colored paper within the SpinnerBot's territory. Program it to bring capture flags back to its "home" and deposit them there.

- Build a basket above the RCX, and remove the three-quarter-length pin from the Spinner Arms sub-assembly that limits the upper range of the arm's motion. Now the SpinnerBot will release a captured flag into the basket. Go flag collecting!

# Robot 10

## RIS Turtle

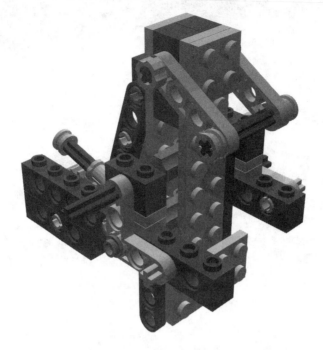

This robot was inspired by the Logo Turtle that was originally developed at the Massachusetts Institute of Technology in the 1960s. Daniel Bobrow and Wallace Feurzeig, Beranek and Newman, Inc., and Seymour Papert developed a programming language called *Logo*, which made computers more accessible to non-experts and even children. Their goal in developing Logo was to allow people to use computers to manipulate more familiar things than numbers and equations. The Logo language borrowed the techniques of symbolic computation (manipulating words and ideas) from the LISP programming language that was used in artificial intelligence research.

When Logo was first developed, it was used to control a simple robot that was called the "Turtle" because the first one had a turtle-like plastic shell. Children would type commands such as FORWARD 50 to make the robot go forward 50 steps, or they would type RIGHT 90 to make it turn right 90 degrees. The turtle robot carried a pen, so children could program the turtle to make drawings on a piece of paper. Additional information on the original Logo Project can be found at www.erzwiss.uni-hamburg.de/Sonstiges/Logo/logofaqx.htm.

Computer programming has been brought to life by means of robotics; robot builders can immediately see the results of their programs. The Logo Turtle is an example of a very powerful teaching aid for programming that reinforced abstract concepts with a real mechanical device. This is also the great strength of the LEGO Robotics Invention System (RIS), which is why this "turtle" is such an appropriate robot to build with the RIS.

In order to create a successful turtle robot, three key features must be in place. First, the robot must be able to drive in straight lines and turn accurately.

The Logo Turtle used precise servo motors driving both wheels. This allowed the speed of the wheels to be accurately controlled so the robot drove straight. With LEGO, the speed of the motors cannot be controlled quite so accurately, so a LEGO robot tends to drive in a wavy line. Our solution to this problem is mechanical: Both wheels on the RIS Turtle are driven from the same motor, so they always drive at the same speed. The way the robot turns is by shifting gears on the right wheel to reverse its direction. When turning, the wheels both turn at the same speed, but in opposite directions.

The second important feature a turtle robot must have is that it must be able to measure distances when driving straight, and measure angles when turning. The Logo Turtle most likely used an optical encoder (rotation sensor) to accomplish this. Because you may not have a LEGO rotation sensor (the RIS does not include one), we will improvise and use the light sensor pointed at a rotating disk to measure distances and angles.

The third feature that a turtle robot must have is the ability to raise and lower its pen. It is fairly straightforward to create a motorized penholder, but it is important to remember that the RIS only includes two motors. One of the motors must be used for driving the robot, and the second motor must be used for shifting the gears to allow turning, so we need a third motor for the pen-holder! Here is where some mechanical ingenuity comes in. You can use one motor to do two different things with some special gearing involving the differential. Basically, the motor performs one function when driving forward and performs the other function when reversing. We use this mechanism to allow the second motor to control both the gear shifter and the penholder. All of this should become apparent when you build the robot.

Once the robot is built, it certainly needs a program to control it. We wrote a NQC (Not Quite C) program that allows users to program the RIS Turtle in a way very similar to the Logo Turtle. The robot will understand commands such as the following:

```
pendown();
forward(30);
right(90);
reverse(10);
```

You can even make the turtle robot "smart" by writing your own function such as *box(x)*, which could draw a box of any desired size:

```
void box(int x)
{
    pendown();
    repeat(4)
    {
        forward(x);
        right(90);
```

```
    }
    penup();
}
```

In order to do this, we have written functions that monitor the rotations of the drive motor by means of the light sensor. These functions translate the driving and turning commands into the appropriate number of motor rotations and then carry out the movements. Due to the nature of the robot's design, it must pause and shift gears prior to turning, which is also defined in the functions. At the very beginning of the program, the robot undergoes an initialization sequence, which brings the gear shifter and penholder into the correct positions. This is necessary because the robot does not know which position these mechanisms are in when it is first turned on. We have included comments throughout the program to explain the key points. Expert programmers can feel free to use our ideas and completely rewrite the program, while beginners should use it "as is" with the basic driving and turning commands.

NQC is a free program that has become very popular amongst LEGO robotics enthusiasts. The best way to use NQC to program your robots is in conjunction with the graphical interface called Bricx Command Center, or BricxCC for short. This can be downloaded from http://members.aol.com/johnbinder/bricxcc.htm. The NQC program for the RIS Turtle is available for download on the Syngress Solutions Web site for the book (www.syngress.com/solutions).

# The Penholder

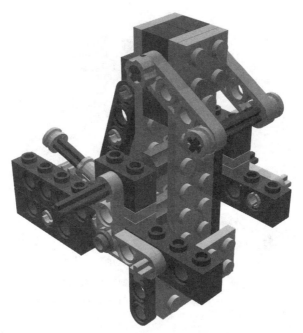

The first sub-assembly is the penholder. The penholder is the mechanism that grips the pen firmly and allows it to move up off the paper and down onto the paper. You may have to adjust the penholder to suit the pen you are using (you can add or remove plates to do this). Ideally, the pen should be as close as possible to the center of the wheelbase. This allows the Turtle to turn while the pen is down without making a mess.

### Penholder Step 0

x1

x1

x1

Start building the side of the penholder.

### Penholder Step 1

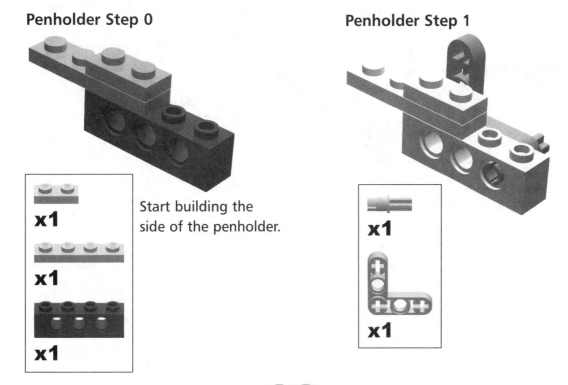

x1

x1

## Penholder Step 2

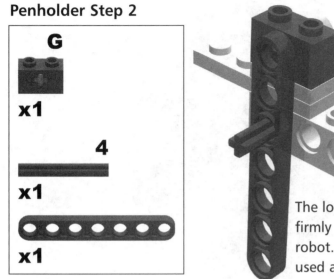

The long liftarm will be used later to firmly connect the penholder to the robot. Please be aware that we have used a green 1x2 brick with an axle hole in this step.

## Penholder Step 3

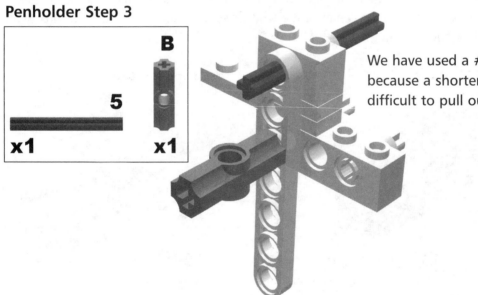

We have used a #5 axle here because a shorter axle is very difficult to pull out!

## Penholder Step 4

**B**

8

x1    x1

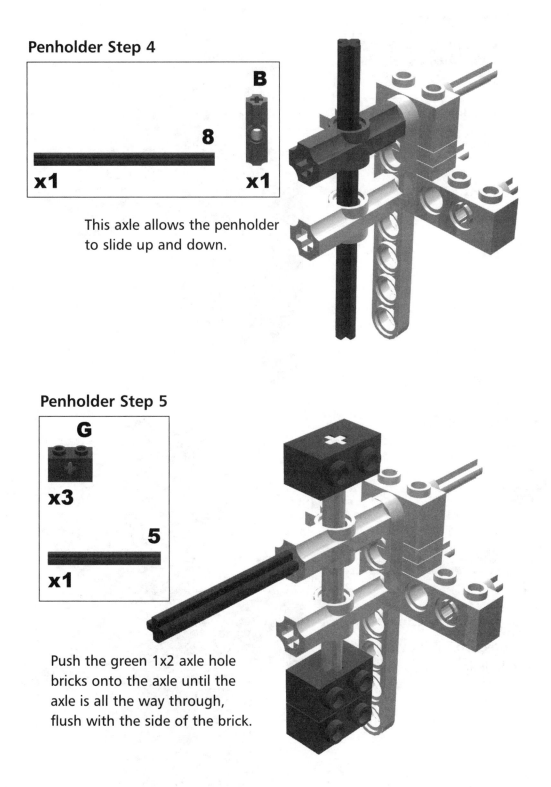

This axle allows the penholder to slide up and down.

## Penholder Step 5

**G**

x3

5

x1

Push the green 1x2 axle hole bricks onto the axle until the axle is all the way through, flush with the side of the brick.

## Penholder Step 6

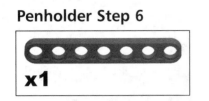

**x1**

One side is complete; now
start building the other side.

## Penholder Step 7

**G**

**x1**

**x1**

Please note that we have
used a green 1x2 brick with
an axle hole in this step.

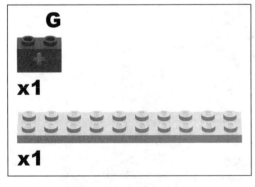

## Penholder Step 8

x1

x1

4

x1

x1

x2

## Penholder Step 9

x1

x1

Y

x1

x1

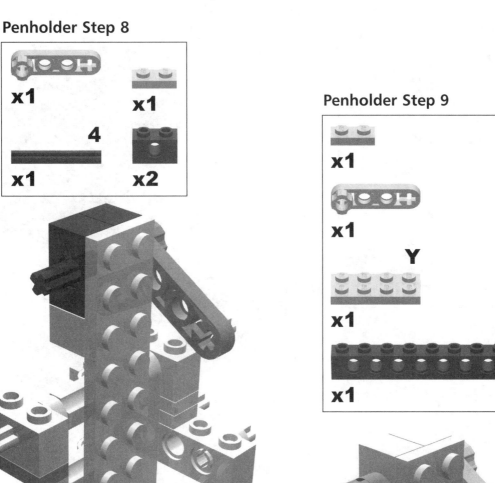

The pen will be held in place against the back plate with the 1x4 liftarms. In this step we have used a yellow 2x4 plate.

## Penholder Step 10

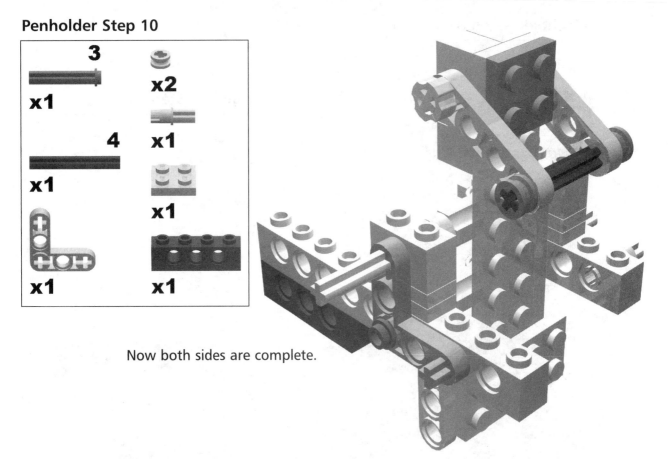

Now both sides are complete.

## Penholder Step 11

Now we will add some parts to the back of the penholder. This gear is used to lift the pen up and down by means of a crank.

## Penholder Step 12

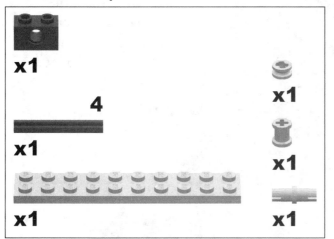

A ratchet is necessary to allow the penholder to run in only one direction. When the motor reverses to shift gears, the ratchet locks the penholder, and it doesn't move. Here the ratchet on the gear is complete.

## Penholder Step 13

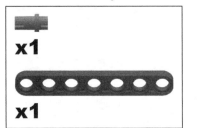

**x1**

**x1**

Add the crank arm. Notice this allows the pen to move up and down while only rotating in one direction.

## Penholder Step 14

**x1**

**x1**

**B**

**x1**

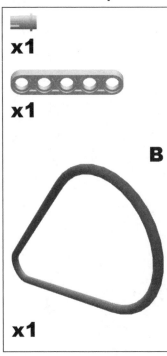

To complete the penholder, wrap a blue rubber band around one half-bushing and one of the 1x4 liftarms, around the back of the penholder, under the 1x2 brick, and then around the half-bushing on the other liftarm.

# The Turtle

Now that the penholder sub-assembly is complete, you can start building the robot. The robot will be built mostly from the bottom up.

## Turtle Step 0

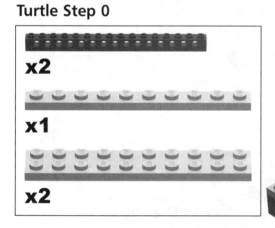

x2

x1

x2

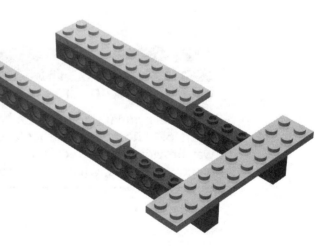

These long beams add support underneath the entire robot.

## Turtle Step 1

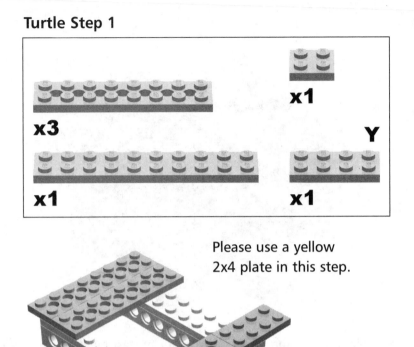

**x3**

**x1**

**Y**

**x1**

**x1**

Please use a yellow
2x4 plate in this step.

Bricks & Chips...

## Fortifying Your Creations with Cross Bracing

Note the sequence here: First a layer of beams, then two
layers of plates, then another layer of beams. Building this
way provides the correct spacing to allow cross bracing
along the beams, which makes for a very strong structure.

## Turtle Step 2

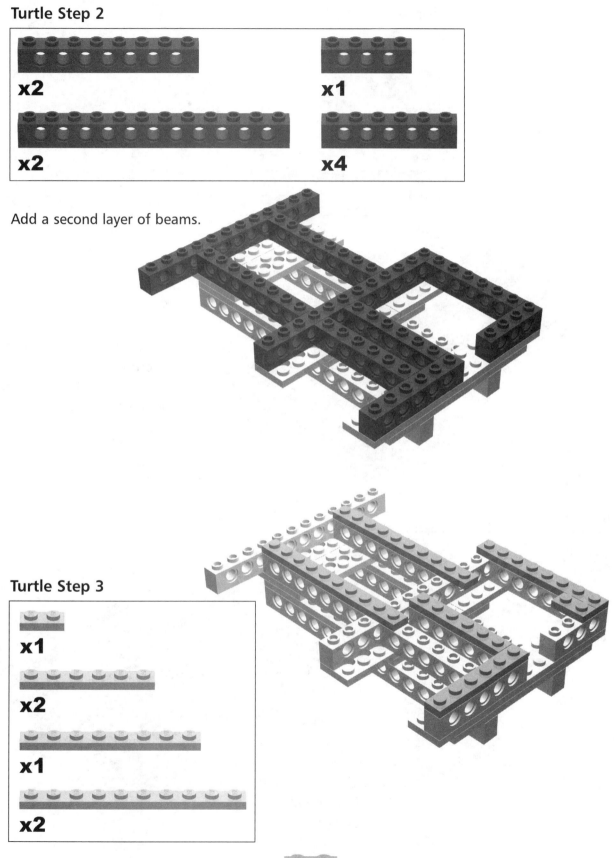

x2

x1

x2

x4

Add a second layer of beams.

## Turtle Step 3

x1

x2

x1

x2

## Turtle Step 4

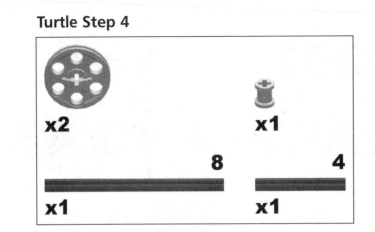

x2      x1

8      4

x1      x1

Add two pulleys. These will be used to transmit power from the motor in front to the back for operating the gear shifter.

## Turtle Step 5

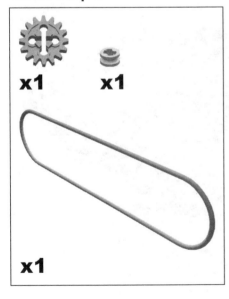

x1      x1

x1

Add a yellow rubber band. You may have to stretch it a bit before you attach it, so it isn't too tight.

## Turtle Step 6

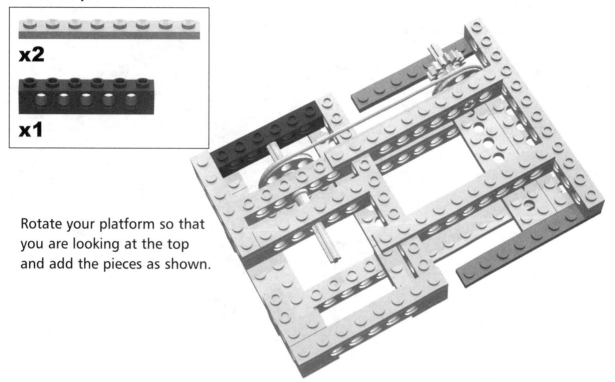

**x2**

**x1**

Rotate your platform so that you are looking at the top and add the pieces as shown.

## Turtle Step 7

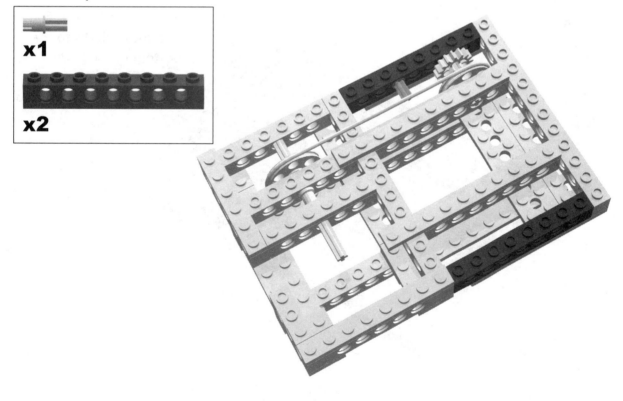

**x1**

**x2**

## Turtle Step 8

**x1**

**x1**

**x1**

Here you add another ratchet to the gear that drives the belt, which in turn will drive the gear shifter. The purpose of this ratchet is the same as the one for the penholder. It only allows the gear shifter to rotate in one direction.

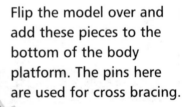

## Turtle Step 9

**x4**

**x1**

Flip the model over and add these pieces to the bottom of the body platform. The pins here are used for cross bracing.

## Turtle Step 10

**B**

**x2**

Please use the blue
2x2 plates in this step.

## Turtle Step 11

**B**

**x2**

Please use the blue 2x2
round plates in this step.

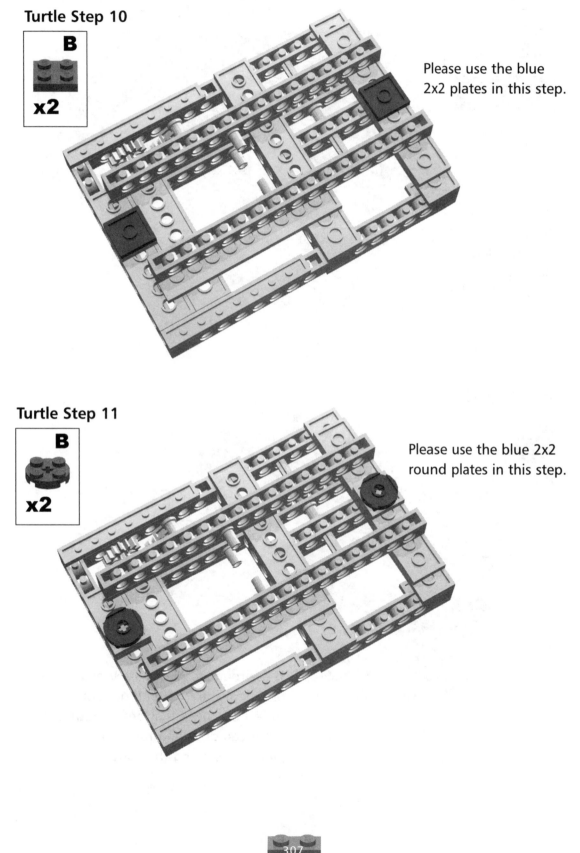

## Turtle Step 12

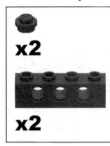

Add the cross brace beams. These hold the turtle platform together on the inside. Later, additional bracing will be added to the outside.

## Turtle Step 13

These rounded pieces are used as skid plates for the front and back of the robot to rest on. Since they are smooth, the robot can drive and turn easily as these plates slide along the paper.

## Turtle Step 14

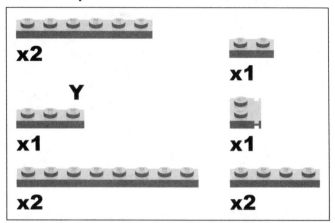

Flip the model back over and add these pieces to the top front portion of the platform.

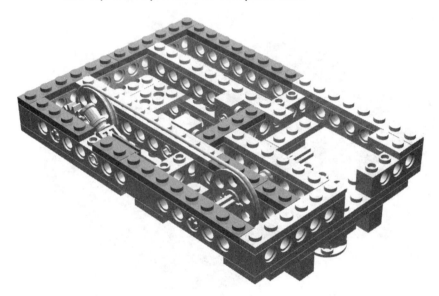

## Turtle Step 15

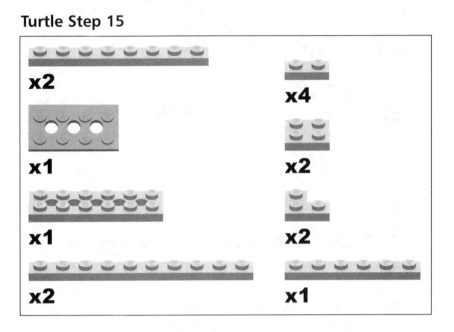

Add another layer of plates.

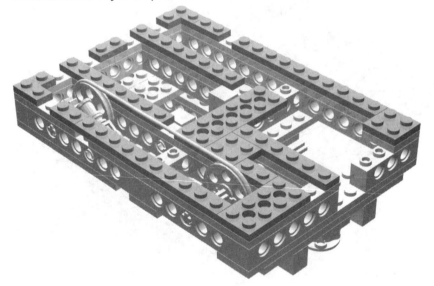

## Turtle Step 16

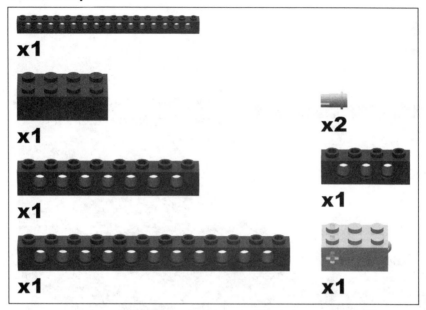

Add another layer of beams and a sensor. This touch sensor will be used to detect the position of the gear shifter.

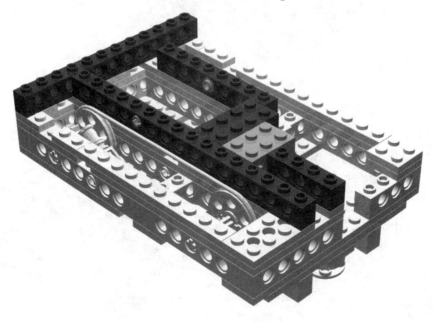

## Turtle Step 17

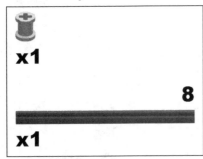

This next sequence of steps shows how the gear shifter is assembled. Follow them very carefully!

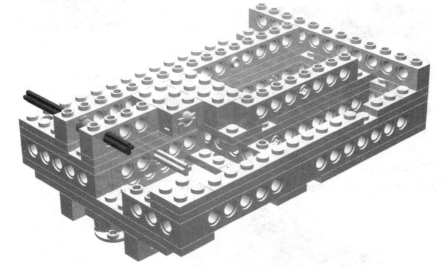

## Turtle Step 18

Add two pulleys to drive the shifter.

## Turtle Step 19

Add two blue belts.

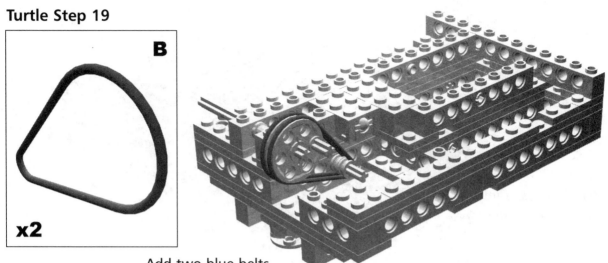

## Turtle Step 20

Place one cam on the pulleys. Note that the axle must be flush with the. The second cam added in the next step cannot have the axle through its center, or the axle used for the crank will not fit.

## Turtle Step 21

Place the next cam on the pulleys. Note that the orientation of these cams is very important: The points must be facing opposite directions!

## Turtle Step 22

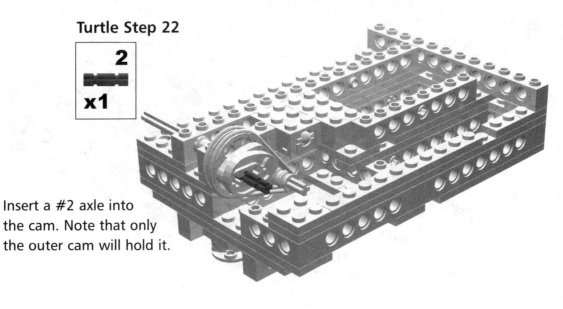

Insert a #2 axle into the cam. Note that only the outer cam will hold it.

## Bricks & Chips...

## Unorthodox Assemblies

The reason for this strange assembly is that a crank with a total stroke of 1 unit is required for the gear shifter. This is because it will eventually be used to shift between two gears that are spaced one unit apart on a shaft. To provide a total stroke of 1 unit, the axle must be connected one half-unit from the drive shaft. There is a special TECHNIC crank piece that is made for a half-unit offset, but the RIS doesn't include any. But two cams can be used together to provide the correct spacing. This is because the cams have several mounting holes at different spacing, and three of these holes just happen to be one half-unit apart. Two cams must be used because the width of an axle is greater than one half-unit, so the axles can't be placed side by side, but must be installed one in front of the other. This is one example of many situations where a construction requires a half unit spacing. The TECHNIC cam is one part that can be used to provide a half unit offset, while other parts include the 1x1 beam with a hole, or 1x2 beam with two holes, or the plate with one bump in the center.

## Turtle Step 23

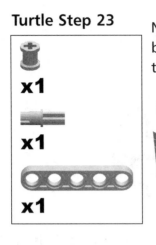

Now add the crank arm. The bushing is used to trigger the touch sensor.

## Turtle Step 24

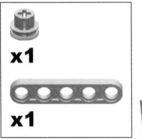

The medium pulley will be used to prevent the #2 axle from sliding out.

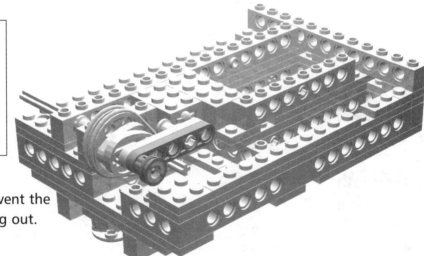

## Turtle Step 25

**3**

This axle is used to move the gear shifter.

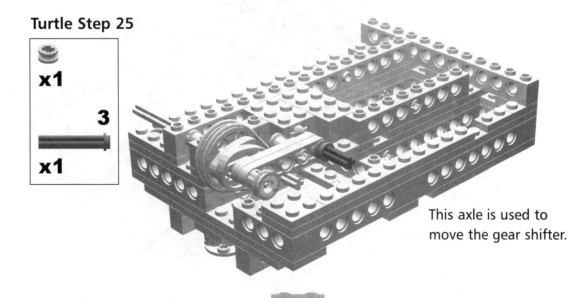

## Turtle Step 26

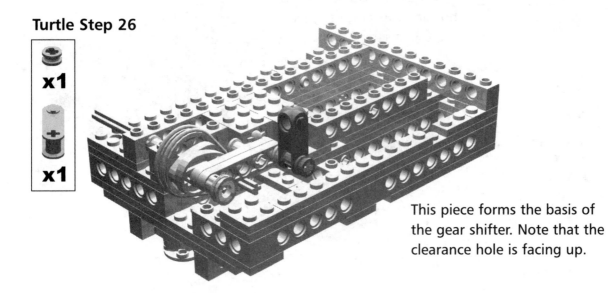

**x1**

**x1**

This piece forms the basis of the gear shifter. Note that the clearance hole is facing up.

## Turtle Step 27

**6**

**x4**

**x1**

**x2**

**x3**

Add the differential gearing to the top front portion of the platform.

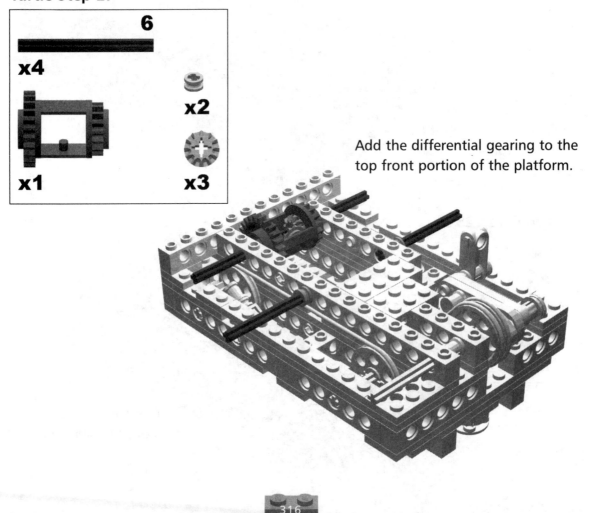

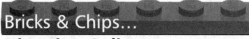

## Direction Splitters

The differential forms the basis of a mechanism we refer to as a *direction splitter*. This is because driving the differential in one direction will perform one task, while driving the differential in the opposite direction performs another task. The functionality is "split" based on the direction in which the motor is turning.

## Turtle Step 28

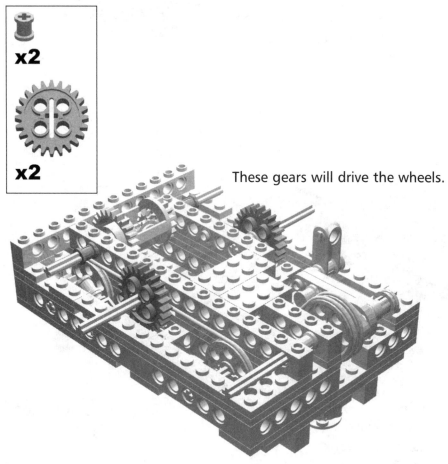

These gears will drive the wheels.

## Turtle Step 29

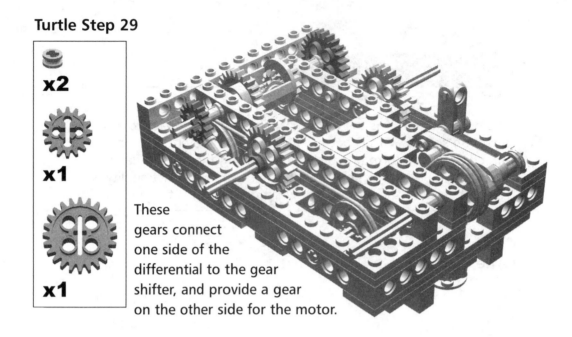

x2

x1

x1

These gears connect one side of the differential to the gear shifter, and provide a gear on the other side for the motor.

## Turtle Step 30

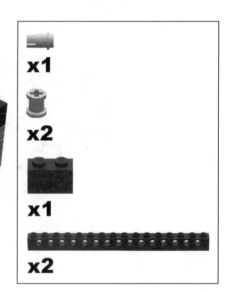

x1

x2

x1

x2

Add a layer of beams. Now you can see the brick that the medium pulley slides against, preventing the gear shifter crank from coming loose. The half-length pin (which is not visible in this step) is inserted in the back beam right behind the gear shifter in the fifth hole from the end. Its purpose is to prevent the gear shifter from catching in the beam's hole at that point.

## Turtle Step 31

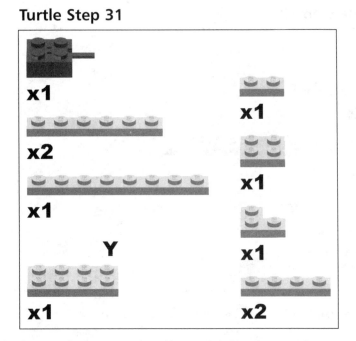

Add a layer of plates and a short wire for the touch sensor.
Note that we have used a yellow 2x4 plate in this step.

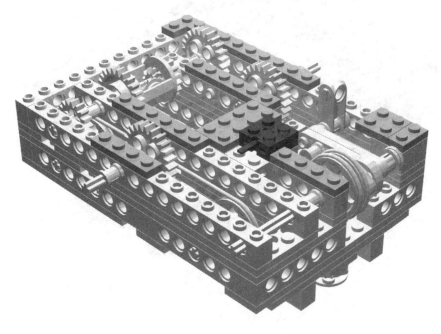

## Turtle Step 32

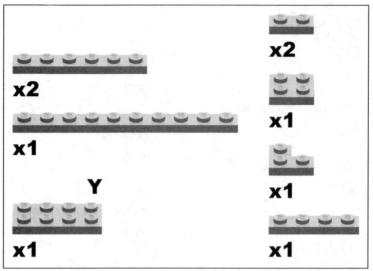

Please use the yellow 2x4 plate for this step.

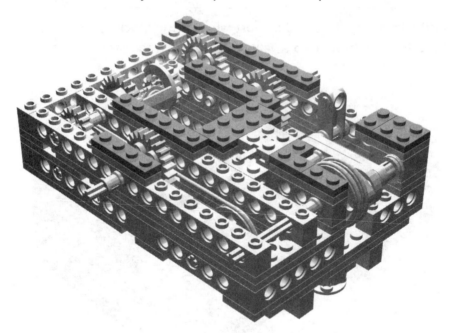

## Turtle Step 33

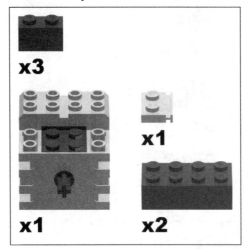

**x3**

**x1**

**x1**

**x2**

Add the drive motor.
The 1x2 plate with a rail is
installed in the bottom groove
on the hidden side of the motor.

## Turtle Step 34

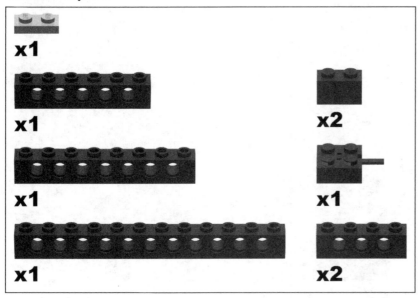

x1

x1

x2

x1

x1

x1

x2

Add a layer of beams and a short wire for the motor.

The touch sensor wire should pass between the motor and the 1x2 brick and above the gear shifter axle.

## Turtle Step 35

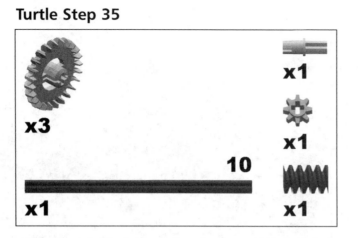

Add the axle and gears for the gear shifter. It is interesting to note the unique property of the worm gear that makes this assembly possible: The axle can slide freely through the worm gear while still able to turn it. The way the gear shifter works is by sliding the two crown gears back and forth along the axle. The direction the axle turns depends on which of the two crown gears is in mesh with the gear that drives them.

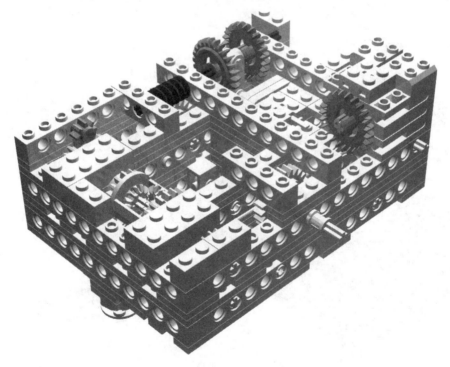

## Turtle Step 36

x1

x1

x1

8

x1

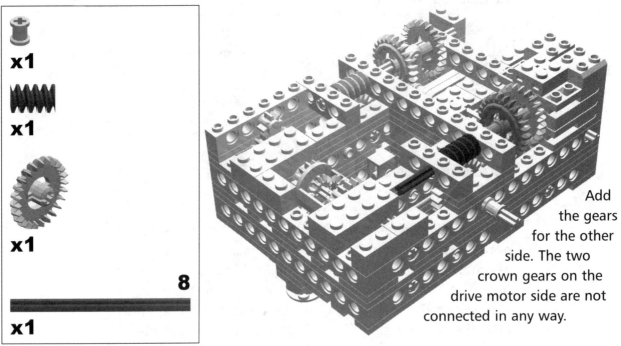

Add the gears for the other side. The two crown gears on the drive motor side are not connected in any way.

## Turtle Step 37

Add the gears and shaft that transmit power from the drive motor to both of the drive axles. The round white plate is used in conjunction with the light sensor to act as a rotation sensor. This will be explained in more detail later.

x1

x2

x1

x1

x2

6

x3

## Turtle Step 38

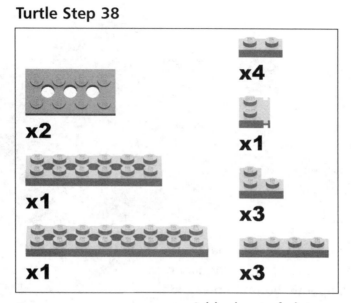

x2

x1

x1

x4

x1

x3

x3

Add a layer of plates.

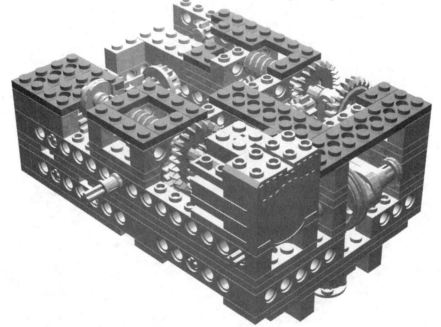

## Turtle Step 39

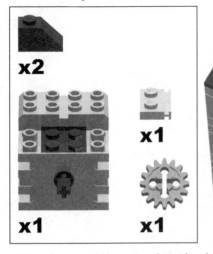

Flip the model around and add these parts to the back of the platform. Now the secondary motor is placed, which will operate both the penholder and the gear shifter.

## Turtle Step 40

Add some plates and a short wire for the motor.

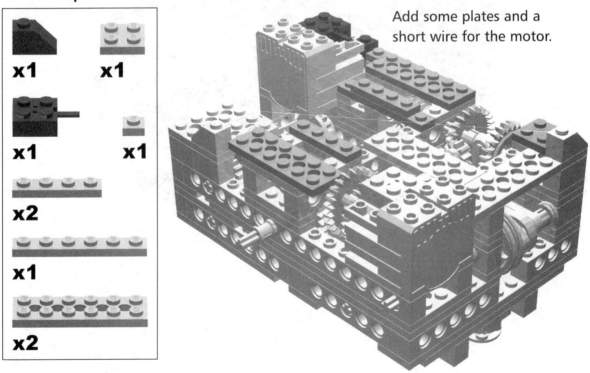

## Turtle Step 41

Add the penholder sub-assembly. Note that the liftarms will have to be pressed inward slightly so they can clip onto the half-pins in the turtle base. This is a bit awkward, but it locks the penholder to the Turtle base, making the assembly more solid. Once the penholder is in place, you can try out the direction splitter mechanism. Turn the secondary motor by hand and watch what happens. When you turn it one way, the gear shifter will move, and when you turn it the other way, the penholder will move. You should also try out the gear shifter at this stage. Turning the secondary motor by hand, shift the gears into one position. Now turn the drive motor by hand and see which way the drive axles turn. Then shift gears into the other position, and turn the drive motor again. The direction the left drive shaft turns should be reversed.

## Turtle Step 42

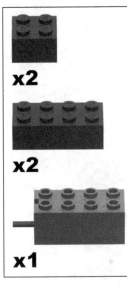

Add the light sensor and some
bricks. The round white 2x2 plate is
used together with the light sensor to
form a rotation sensor. As the plate rotates,
the sensor is either blocked by the bumps or
exposed, creating a transition in the light sensor value.
This transition can be monitored in the program, allowing the robot
to measure driving distances and turning angles. The round plate
must be close to the light sensor, but be sure it isn't touching!

## Turtle Step 43

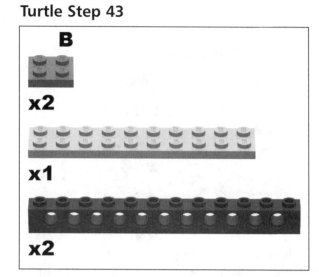

**B**

x2

x1

x2

These beams will serve as a support for the RCX and also to strengthen the robot with cross bracing.

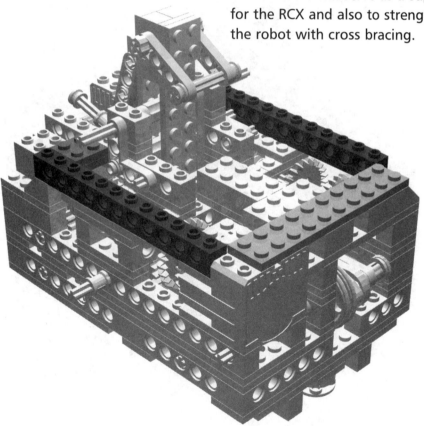

## Turtle Step 44

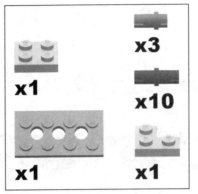

Flip the model around and add the pins for bracing.

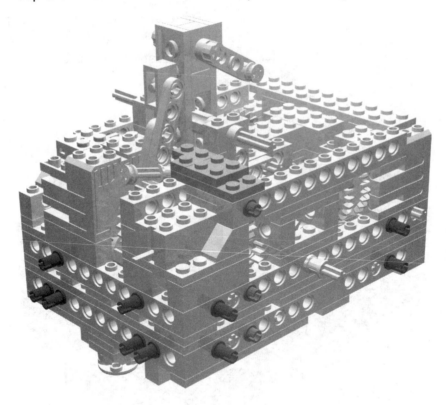

## Turtle Step 45

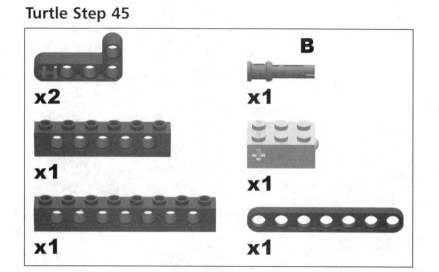

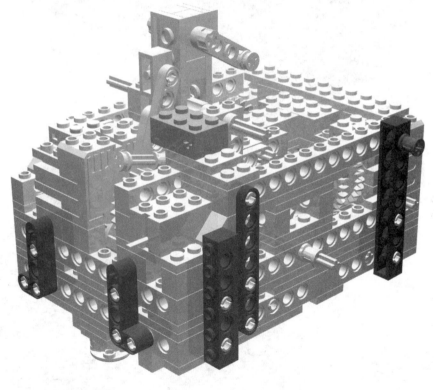

Add the cross bracing beams.

## Turtle Step 46

x3

x5

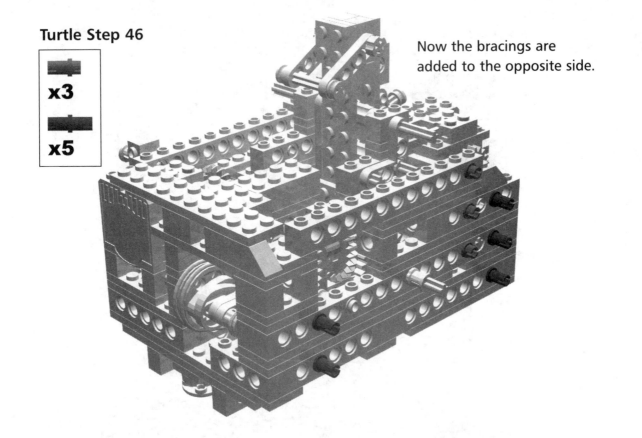

Now the bracings are
added to the opposite side.

## Turtle Step 47

B

x1

x1

x1

x1

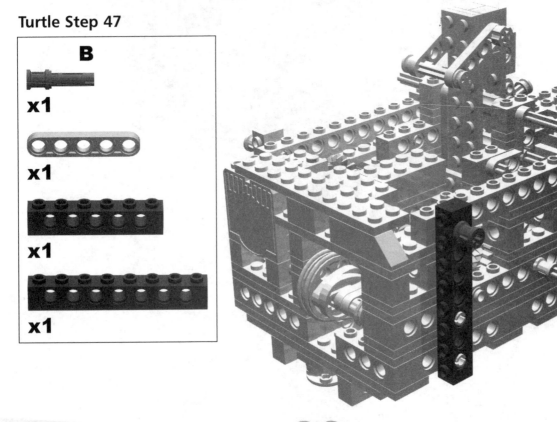

## Turtle Step 48

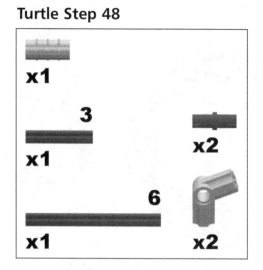

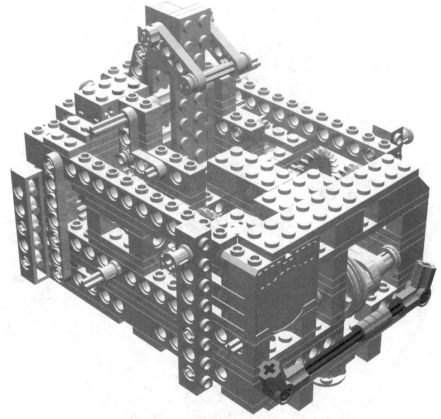

This bumper is merely decorative. Connect two of the black flexible hoses to each other, and run them from the axle sticking out of the side of the penholder down to the connector on the bumper.

## Turtle Step 49

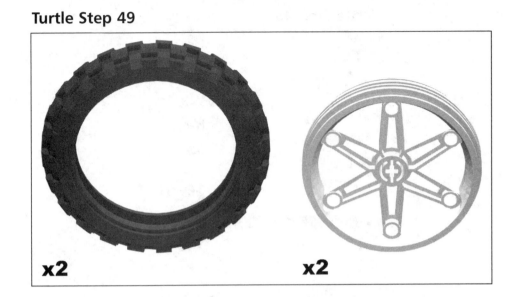

**x2**  **x2**

Finally, add
the wheels.
The Turtle is
slightly under-
powered with only
one motor driving it, so
it is best to run it on flat, smooth sur-
faces. It may seem like a disadvantage that the robot is only driven by one motor,
but in this case it is a huge advantage. Since the same motor drives both wheels,
the wheels will rotate at the same speed, enabling the robot to drive in perfectly
straight lines. If two separate motors were used on each wheel, they would likely
rotate at slightly different speeds, causing the robot to draw crooked lines.

## Turtle Step 50

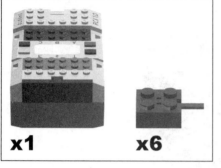

**x1**  **x6**

Add the RCX and then connect all of the wiring. Ensure that the motor wires have the same orientation as shown in the picture. These are the connections:

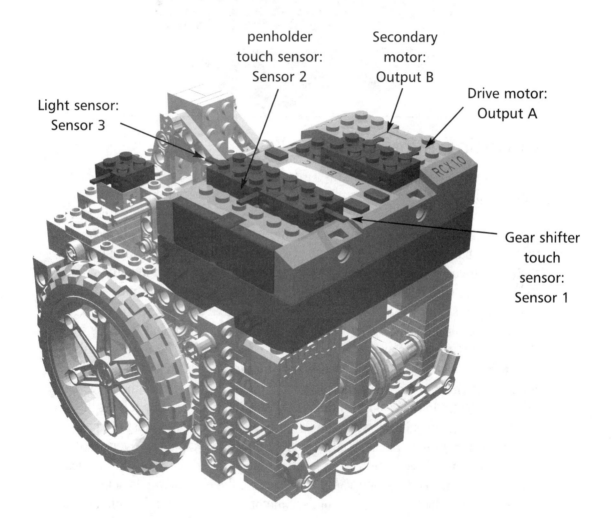

penholder
touch sensor:
Sensor 2

Secondary
motor:
Output B

Drive motor:
Output A

Light sensor:
Sensor 3

Gear shifter
touch
sensor:
Sensor 1

## Choosing a Pen

Now that you are done building the Turtle, you need to choose a pen to use
with it. Felt tip pens or markers make the best pictures, because the robot
doesn't press a pencil or ballpoint pen against the paper hard enough. I used a
Staedler felt tip pen designed for overhead projectors. Sharpie markers also
work well, but they are shorter than the Staedler, so they can be held only at
their end. If you want to use the Sharpie marker, you need to remove the gray
2x2 plate and the yellow 2x4 plate from the penholder. The unfortunate thing
about markers is they tend to bleed a lot, making large dots where the robot
turns. You could prevent this by lifting the pen before turning, but then the
lines may not connect as well. Once you have chosen a pen, you need to make
sure it fits properly in the penholder. When you put the pen in, turn the Turtle
around and check where the pen tip comes through. To make nice drawings,
you want the tip to be as close to the center of turning as possible. Check this
by seeing if the tip is in line with the two drive axles. You can adjust the position of the pen in the holder by either adding or removing plates. If you want
to be overly precise, you could wrap layers of tape around the pen until it lines
up exactly in the center. When you put the pen in, the easiest way to set the
correct height is by turning the secondary motor by hand until the penholder is
in the lowest position. Then you can gently push the pen down until it touches
the paper.

## Writing Your Program

After you have downloaded the NQC program, look through it to get an idea of how it works. We have indicated where to insert your commands in the main task. You can try drawing a square first:

```
Repeat(4)
{
    forward(20);
    right(90);
}
```

Try this out and see what happens! If the Turtle moves but does not stop, you need to adjust the threshold for the light sensor. This is near the beginning of the program, and we have explained how you should set your threshold there. (If you are an experienced programmer, you may want to have the Turtle initialize its own threshold value.) If the Turtle drives backwards when it should be moving forwards, check that the drive motor is wired correctly. Also, if the gears shift when the pen should be lowering, you know you have the secondary motor wired wrong. The orientation of the sensor wires does not matter, just as long as they are connected to the right port. Once you have the Turtle working correctly, try to draw some more shapes–you can even get the Turtle to write your name!

## Things to Watch Out For

Over time, the crown gears on the gear shifter will work their way loose. This can result in poor meshing, and the robot will not drive properly. It is obvious when this happens because you hear a nasty crunching sound as the gears skip. Make sure you check periodically that these gears are pressed firmly together. This is most easily done from the side after removing the wheel. If this happens often to your robot, you could try to keep the gears on the shaft more firmly by means of some rubber cement or other removable glue. Whatever method you try, make sure that the shaft can still rotate freely within the gear shifter piece.

Another problem can arise when the ratchet on the penholder gear jumps. This sometimes happens when the robot is shifting gears for a turn, and the pen also lifts a bit. In order to remedy this, you could either extend the axle on the penholder ratchet to make it heavier or adjust the gear shift sequence in the program. This would involve starting the secondary motor at a lower power, and then accelerate up to full power to prevent the jumping.

# SYNGRESS SOLUTIONS...

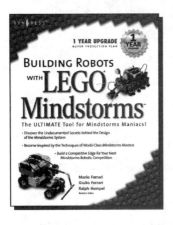

## Building Robots with LEGO MINDSTORMS

The LEGO MINDSTORMS Robotics Invention System (RIS) has been called "the most creative play system ever developed." This book unleashes the full power and potential of the tools, bricks, and components that make up LEGO MINDSTORMS. Some of the world's leading LEGO MINDSTORMS inventors share their knowledge and development secrets. You will discover an incredible range of ideas to inspire your next invention. This is the ultimate insider's look at LEGO MINDSTORMS and is the perfect book whether you build world-class competitive robots or just like to mess around for the fun of it.

ISBN: 1-928994-67-9

Price: $29.95 US, $46.95 CAN

## More Great Books in the Syngress 10 Cool Series!

The 10 Cool Series covers the most popular MINDSTORMS kits from LEGO and these books give you everything you need to create cool robotics projects in under one hour.

### 10 Cool LEGO MINDSTORMS Dark Side Robots, Transports, and Creatures

ISBN: 1-931836-59-0

Price: $24.95 US, $38.95 CAN

### 10 Cool LEGO MINDSTORMS Ultimate Builders Projects

ISBN: 1-931836-60-4

Price: $24.95 US, $38.95 CAN

solutions@syngress.com

SYNGRESS®